Stem Cell Research

Stem Cell Research takes a multidisciplinary approach to the topic of human embryonic stem cell research, starting with the breakthrough discovery up through the present day controversy. The book invites the reader to join the conversation by providing a well-balanced approach to many of the issues surrounding the development of this controversial scientific field. It includes the thoughts and experiences of scientists, journalists, and ethicists, as they have tried to approach the topic through a variety of different academic disciplines.

The book will help the nonscientist understand the biology, research regulations, and funding; and simultaneously it will help the scientist better comprehend the full spectrum of ethical, religious, and policy debates.

Toni Marzotto is Professor of Political Science at Towson University. She teaches courses in American government and public policy.

Patricia M. Alt is Professor of Interprofessional Health Studies at Towson University. She teaches courses in public policy, ethical issues in health care, long-term care, and the American health care system.

Stem Cell Research

Hope or Hype?

Toni Marzotto and Patricia M. Alt

London and New York

First published 2017
by Routledge
2 Park Square, Milton Park, Abingdon, Oxon OX14 4RN

and by Routledge
711 Third Avenue, New York, NY 10017

Routledge is an imprint of the Taylor & Francis Group, an informa business

British Library Cataloguing-in-Publication Data
A catalogue record for this book is available from the British Library

Library of Congress Cataloging in Publication Data
Names: Marzotto, Toni, 1945- author. | Alt, Patricia Maloney, author.
Title: Stem cell research : hope or hype? / Toni Marzotto and Patricia M. Alt.
Description: Milton Park, Abingdon, Oxon ; New York, NY : Routledge, 2017. | Includes bibliographical references and index.
Identifiers: LCCN 2017004556| ISBN 9781498756549 (hbk) | ISBN 9781138627291 (pbk) | ISBN 9781498756556 (ebk)
Subjects: LCSH: Stem cells. | Stem cells--Research. | Stem cells--Moral and ethical aspects.
Classification: LCC QH588.S83 M374 2017 | DDC 616.02/774--dc23
LC record available at https://lccn.loc.gov/2017004556

ISBN: 978-1-4987-5654-9 (hbk)
ISBN: 978-1-138-62729-1 (pbk)
ISBN: 978-1-4987-5655-6 (ebk)

Typeset in Times New Roman PS Std
by Nova Techset Private Limited, Bengaluru & Chennai, India

Contents

List of figures

List of tables

Preface

This book had its genesis in 2006, when both authors became interested in the role of the media and legislators in promoting the passage of the Stem Cell Research Act of 2006 in Maryland. Our research subsequently expanded to include other states and the role played by the media, particularly during the Bush Administration, in framing the debate. The focus expanded again in 2009 and 2010, when the Obama Administration extended federal funding for additional human embryonic stem cell lines. By this time, a number of states had jumped on the bandwagon and were actively promoting and funding stem cell research.

The idea of an interdisciplinary text emerged when Towson University (Maryland) adopted a new core curriculum that promoted critical thinking and writing to improve student analytical skills. This topic became a perfect vehicle through which to explore the various disciplines that were discussing human embryonic stem cell research from their own perspectives. The student activities at the end of the substantive chapters are a way to help our readers think about the issues as viewed by the various disciplines. Our motto is: think like a scientist, or a journalist, or an economist, and so forth. That task is more complicated than it initially appears to be.

We would like to thank Linda Caplis, professor of the Towson Seminar course on Stem Cells, for her encouragement and assistance as we moved toward the development of this text. A number of graduate students, especially Kristin Babin, Meghan Dilly, and Tyler New, played important roles in moving the project forward. We also thank the Towson Faculty Development and Research Committee for providing funding in the initial stages of this work.

Toni Marzotto would like to thank Towson University and the Department of Political Science for granting her a yearlong sabbatical to work on the manuscript. The year permitted her to follow her intellectual curiosity and to learn more about the science of stem cells. A special thanks to Suzanne Legault and Neal Harwood, who read and edited chapters; sent her articles; and challenged her throughout the process. She would also like to thank Chris and Carla Jackson, who could not wait for their mother to finish the manuscript, so that the family could talk about something other than stem cells.

Patricia M. Alt would like to thank Toni Marzotto and Linda Caplis for their encouragement as we moved ahead on this project, and especially Toni, as she has carried the majority load over the last year.

About the book

Stem cell research is more than just the science of how these cells are derived and their potential. The research, initially hailed as a major discovery, was soon caught up in the sort of moral dilemma that is not uncommon in the scientific world. The use of in vitro fertilization (IVF) or animal cloning, and now gene-editing discoveries, are enveloped in controversy. Scientists playing God in their labs is a frequent thread that is heard about many scientific discoveries. In some cases the controversy subsides; in others it does not.

This book looks at stem cell research from the perspectives of the various disciplines that have produced extensive writings on the topic. It is organized to help readers appreciate the complexity of issues that become matters of public policy and public discourse. The science is important, but the reactions of legislators, journalists, interest groups, and the public play a very important role in supporting the development of the science. The subtext of our book is about bringing this science into the public arena, and what happens when the science is controversial.

Toni Marzotto is a Professor in the Political Science Department at Towson University.

Patricia M. Alt is a Professor in the Department of Interprofessional Health Studies at Towson University.

1 Introduction

The path forward

> The issue of research involving stem cells derived from human embryos is increasingly the subject of a national debate and dinner table discussions.... Many people are finding out that the more they know about stem cell research, the less certain they are about the right ethical and moral conclusions.
>
> —President George W. Bush, August 9, 2001[1]

> At this moment, the full promise of stem cell research remains unknown, and it should not be overstated. But scientists believe these tiny cells may have the potential to help us understand, and possibly cure, some of our most devastating diseases and conditions. To regenerate a severed spinal cord and lift someone from a wheelchair. To spur insulin production and spare a child from a lifetime of needles. To treat Parkinson's, cancer, heart disease and others that affect millions of Americans and the people who love them.
>
> —President Barack Obama, March 9, 2009[2]

The above quotes attest to the importance of stem cell research. They also confirm that the much-anticipated breakthroughs are still on the horizon as research continues in this developing field. The field of regenerative medicine has become a multidisciplinary billion-dollar global industry akin to the dot-com boom of the 1990s. The field holds the promise of improving the physical well-being of future generations. It may not offer the miracle cures that some expected, but the hope is that it will end the pain and suffering of those with disabling diseases. Our book is about this research and the controversy that accompanies it. The subtitle of this text, Hope or Hype, was chosen to explore the dual nature of this expanding field of research. Are we on the brink of finding cures for some of the most debilitating diseases affecting humans, or are we on an elusive search for a mythical fountain of youth? You be the judge after reading this book.

Stem cells are cells with the ability to divide indefinitely so that they can both replenish and renew themselves. They can give rise to specialized or differentiated cells that are assigned a specific job depending on the tissue in which the cells are located. Stem cells contribute to the body's ability to renew and repair its tissues. Out of the more than 100 trillion cells in the human body less than one percent are stem cells.[3]

Few people had heard about stem cells before 1998, when two research teams published papers reporting that they had isolated human embryonic stem cells (hESCs). Both papers were published in scholarly scientific journals within weeks of each other.[4] An article in the journal *Science* lauded the work of both teams as a major breakthrough in the field of cell biology. The author, writing in 1998, speculates that, "In a matter of years, some researchers say, it may even be possible to use such cells to repair blood, bone, and other tissues."[5] The media had yet to learn about the scientific breakthrough; a survey of articles in national newspapers from 1999 to 2000 reveals fewer than 25 stories with "stem cells" in their titles. By 2001, however, the number had doubled. The year 2001 was a formative one for the American public's awareness and perception of this issue. On August 9, 2001, President George W. Bush delivered a televised speech to the nation (the first of his presidency) on the topic of stem cell research.[1] He followed up, 2 days later, by devoting his weekly radio address to the same topic.[6] While he referred to the great hope for treatment, he also relayed his concern that the research destroys the potential for life. The dichotomy could not have been clearer, and the national conversation began.

As if on cue, 2 weeks later *Time* magazine featured James Thomson (a stem cell researcher from one of the 1998 teams) on its cover with the caption "The Man Who Brought You Stem Cells." After Thomson was interviewed about his landmark discovery, he was asked what was next on his agenda. Thomson replied that he wanted to get back to work and obscurity.[7] However, the genie was out of the bottle.

The coming decade would see the field expand as scientists in the United States and across the globe searched for ways to harness human embryonic and induced pluripotent stem cells (iPSCs, discovered in 2007) and apply them to clinical therapies. With the exception of a few small clinical trials (discussed in Chapter 2), the search continues. Today, human embryonic stem cell research crosses many disciplines, including reproductive biology, immunology, oncology, transplantation, ethics, and policy. With many research teams working with these cells, it is likely that breakthroughs will emerge in the foreseeable future.

The objective of this text is threefold. First, given the multidisciplinary nature of the research, the text invites the reader to join the conversation. Understanding the science basics of stem cell research is an important and necessary first step, one that allows the reader to better understand the debate over government regulations, the controversy over research ethics, the funding of research, and, ultimately, who will benefit from the research. Scientific research, especially in the life sciences, is likely to have ethical ramifications, monetary implications, and safety concerns; this could not be more relevant to the field of stem cell research. Scientific inquiry, as with most fields of research, does not happen in a vacuum; it is affected by many overlapping forces. Understanding these potential forces is our first important objective.

Second, becoming a critical thinker requires that the reader go beyond the chapters in the text. We have synthesized the key issues within each chapter, and each chapter also includes a list of scholarly articles that will give the reader an opportunity to see how the experts present or frame their research, be they biologists, ethicists, or journalists. Articles in popular journals and magazines provide a context for understanding the public's exposure to the issues. The text is a jumping-off point for expanding your critical thinking about one or more of the subtopics in this book.

Third, critical writing follows critical thinking. Each chapter asks the reader to articulate his or her point of view on a specific topic. Some of these topics are polemical, and we ask the reader to develop a thesis and follow it through. Supporting a thesis based on reliable sources is an important step in the development of clear, coherent, and persuasive writing.

The following synopsis provides an overview of each chapter's contents.

Chapter 2—Joining the Conversation: What Are Stem Cells? And Why Should You Care?

This chapter begins by defining the three stem cell types. The scientific and popular literature uses different acronyms for embryonic stem cells. We will use hESC for human embryonic stem cells, iPSC for induced pluripotent stem cells, and AS for adult stem cells.

Although hESCs continue to be the most controversial, the discovery of iPSCs in 2007 did not end the debate.[8] As we will discuss, research findings suggest that iPSCs are problematic and are not as effective as hESCs at developing into all cell types.[9] Diagrams will give the reader a visual overview of how scientists extract and grow these cell types in vitro.

The chapter will cycle back to a brief overview of how stem cell science emerged, and the controversies that affected these early developments. Without the process of in vitro fertilization (IVF), there would be no excess embryos from which to grow stem cell lines. In 1996, the birth of Dolly, the infamous cloned sheep, caused an uproar, fueled by the fear that humans would be next.[10] This has not occurred, but as far as we can tell most animals have been cloned (mostly for commercial reasons).

Dr. James Thomson and Dr. John Gearhart's studies isolating and growing the first human embryonic stem cell lines are discussed, followed by the work of Shinya Yamanaka, who combined genetics with cell biology to create iPSCs. Despite an exponential growth in the field, human clinical trials are still few and far between. However, stem cells are contributing to the study of diseases through the testing of drugs on cells in Petri dishes.

Chapter 3—Stem Cell Federalism: A Legal and Regulatory Quilt

Since the end of World War II, the federal government has provided much of the financial support for basic scientific and medical research. However, the government also recognizes the need to regulate scientific research, especially research conducted on human subjects.

Chapter 3 looks at the laws, regulations, and guidance that have affected the development of stem cell research at the federal level and, later, at the state level.

Between 1973 and 1993, no federal funding was available for embryonic stem cell research. In 1993 Congress enacted the National Institutes of Health (NIH) Revitalization Act, giving the NIH the authorization to support human embryo research.[11] However, President Bill Clinton directed the NIH to forego funding for any projects involving the creation of embryos solely for research purposes.[12]

Two events occurred in 1996 that reshaped federal funding. First, in 1994 the Republican party became the majority party in both houses of Congress. President Clinton did not wade into these controversial issues with the new Republican majority in Congress. Second, Congress enacted the Dickey–Wicker Amendment that prohibited the use of federal funds for human embryonic research.[13]

President Bush established his own policy regarding stem cell research. He prohibited the use of federal funds for:

1. Stem cells derived from embryos destroyed after August 9, 2001, and
2. The creation of human embryos for research purposes.

All the stem cell lines in existence prior to 2001 were created with private funds, so growing the lines or keeping them alive would not lead to additional destruction. Researchers using private funds could continue to derive new lines after 2001, but these lines were not eligible for federal funding. Bush's limitations on funding led to continuous efforts by the then-majority Democratic Congress to increase funding, and pushed stem cell research onto the agenda of state governments. By 2004, a number of states with thriving biomedical sectors moved to fund hESC research. California, New York, and Massachusetts were among the first to move in this direction. California, in a bid to become the stem cell state, enacted a 10-year, $3 billion initiative.

The remainder of this chapter discusses the states that used their own funds to support research. States hoped that the investment would lead to job creation as well as an improved tax base, as both new companies and new employees moved to the state. By 2015 we began to see some return on investment as clinical trials were announced and new findings published.[14]

Chapter 4—Of Facts and Frames: Sharing the News and Influencing Views

The role of the media in a democratic society is to bring issues to the public's attention. One theory is that the media is a conduit—a vehicle that brings new information to the general public in language that can be easily understood.[15]

But the media can also be a conductor, shining a flashlight on an issue or event and renewing interest long after public interest has waned. The media can present information in such a way as to create a positive or negative public view.[16]

In this chapter, we look at the role played by the media in disseminating information about stem cell research before, during, and after Bush issued his limitation on stem cell research. The Obama Administration's expansion of stem cell funding in 2009 was met with less media attention. By 2009, there were robust state and private funding sources that provided scientists with resources to create their own stem cell lines.[17]

It is fair to say that few in the media had heard about the potential for embryonic stem cell research prior to Bush's speech. News coverage spiked after his speech and remained high as the president battled with a Democratic Congress trying to expand funding. We look at articles in the press and scholarly articles assessing the role of the press in framing their stories. In the process of informing and communicating, the media can influence public views and attitudes about controversial issues. Numerous opinion polls emerged shortly after 2001 in which trend analyses done by large polling organizations showed the remarkable stability of public attitudes about stem cell research—the majority approved. Finally, there was a growing public relations effort, especially by researchers, to publicize their findings and to discuss the potential for curing diseases. The increase in the number of TED Talks, interviews with journalists, scientific lectures mounted on YouTube, and so forth, were all evidence that the scientific community was becoming adept at using the Internet and social media to disseminate their research.

Chapter 5—Costs and Consequences: Funding Fragmentation

It is hard to conduct medical research without money. The NIH has always been a major source of funding for basic medical research, but in the area of embryonic stem cells this was not the case.[18] Geron Corporation, a private company, funded the research of both Thomson and Gearhart and became the first to use hESC in clinical trials.[19]

The hiatus between the isolation of embryonic stem cell lines and Bush's approval for funding of a limited number of stem cell lines led states to look for alternatives. States with established and well-funded biotechnology sectors moved to fill the void in federal funding.

This chapter provides an overview of the changes in NIH funding of stem cell research after 2002 and through 2016. The NIH continues to be a major source of basic science funding, but the passage of Proposition 71 in California turned that state into a major player in the field, and the state's role in expanding the field is examined.

The second part of this chapter examines the nonmonetary consequences of stem cell research. The prestige of the U.S. as a global leader in biotechnology was confirmed when Thomson and Gearhart became the first to isolate hESC. But that position was then challenged when the U.S. limited funding while other countries vied to become global stem cell hubs.[17]

Chapter 6—Ethical Dilemmas: Always Changing

No book on stem cell research would be complete without a chapter on the ethical and moral issues that continue to swirl around the field.[20] Although the heated controversy seems to have subsided (our theory), the issue could reemerge. It is important for the reader to understand how individuals and groups continue to hold dichotomous views of this research.

The major ethical issue that continues to emerge is based on the question: When does life begin? Abortion is legal in the United States, although a significant portion of the population believes it should not be.[21] There are regular efforts to get either Congress or the Supreme Court to declare abortion illegal. This is linked to the issue of using embryos as sources of stem cell lines.

Religious views, as well as ethical and moral concerns, continue to play a role in defining what ought to be permissible in stem cell research. In some cases, stem cell research opponents manifest these concerns as portrayals of scientists experimenting with human subjects.

The need for protection of human subjects led to the creation of a number of government panels and commissions to ensure that scientists secure the informed consent of their patients, especially in clinical trials. These delays often annoy patient advocacy groups, as well as private companies eager to test new therapies, but moving from bench to bedside is a necessary first step that must be taken with caution. After all, humans are not lab mice. So, at what point do these clinical trials ensure safety? Some trials listed on the NIH webpage are being carried out in foreign countries where laws are more relaxed. Does the NIH have an obligation to ensure that the clinical trials they list are safe, even if they do not provide financial support for them? The issues of fraud in scientific research and endangerment of patients in clinical trials are not mutually exclusive; there is some concern that scientists and physicians might engage in trials that injure patients in their rush to be first.

This chapter also examines the economic consequences of future treatments. Who will be able to get the new therapies? How much will they cost? Who will pay? There is some speculation that therapies will cost hundreds of thousands of dollars.[22] At the experimental stage, companies and nonprofit institutions are willing to donate their time and money to see if a therapy works. Getting the discovery on the front page of a major newspaper or on the evening news is great publicity, but it also creates an expectation that the therapy will be available to all who need it and that health insurance plans will help cover the costs. However, stem cell therapies will not be mass-produced like prescription drugs and will require individualized administration.

Finally, who owns the technology? As stem cell treatments move from bench to bedside the transition will be done by private companies that are founded on a for-profit motive. Supporting a clinical trial is the first step in getting a treatment approved for commercial development. We welcome companies footing the bill for clinical trials; will we be as welcoming when we see the market price of the final product?

Ethics and the use of embryonic stem cells are not solely tied to religious and moral beliefs, but also to future issues of cost and access. Life is better if you can see clearly, but macular degeneration is not a life-threatening condition. Are drug companies expected to provide compassionate care at a reduced cost? You decide.

Chapter 7—Conclusion: In Our Lifetime?

Stem cell research no longer makes the front-page news. It is not the topic of contentious legislative hearings, proposed bills, or presidential vetoes, but it continues to be a subject of intense scientific investigation. More scientists have entered the field, joined by professionals in a myriad of other related fields. The institutes of regenerative medicine employ the likes of engineers, computer programmers, social workers, psychiatrists, cell biologists, and many others. More money is being spent by the federal government, state governments, and the private sector, and translation of this technology to actual treatments is on the minds of all concerned.

It is likely that these long-awaited treatments will take a bit more time to realize. The first bone marrow transplant took place in 1956; today, over 20,000 are performed annually. We have yet to see a successful clinical trial using hESC or iPSC, although a number are being administered. That successful first trial will be the opening bell in a race to find new ways to use these therapies.

Notes

1. G. W. Bush. Address to the nation. Stem cell research. *Public Papers of the Presidents of the United States*, Book 02, July 1–December 31, 2001, August 9, 2001: 954.
2. B. Obama. Remarks on signing an executive order removing barriers to responsible scientific research involving human stem cells and a memorandum on scientific integrity, *Public Papers of the Presidents of the United States*, Book 01, March 9, 2009: 199.
3. J. Slack. *Stem Cells: A Very Short Introduction*, Oxford: Oxford University Press, 2012.
4. J. A. Thomson et al., Embryonic stem cell lines derived from human blastocysts, *Science* 282, November 1998: 1145–1147; M. J. Shamblott et al., Derivation of pluripotent stem cells from cultured human primordial germ cells, *Proceedings of the National Academy of Sciences* 95, November 1998: 13726–13731. Both research teams were privately funded by the Geron Corporation, based in Menlo Park, California. This eliminated any potential conflict over federal funding of stem cell research. The California-based biotechnology company was started in 1990 by Michael West, an entrepreneurial scientist with a PhD in cell biology. He went on to form the company BioTime, Inc., that today include a number of subsidiaries, one of which is Asterias Biotherapeutics, Inc., that recently receive approval from the FDA to begin trials on patients with spinal cord injuries. The clinical trial is funded by a grant from the California Institute of Regenerative Medicine.

5. E. Marshall, A versatile cell line raises scientific hopes, legal questions, *Science* 282, November 6, 1998: 1014–1015.
6. G. W. Bush. The president's radio address, August 11, 2001, *Public Papers of the Presidents of the United States*, Book 02, August 11, 2001: 956.
7. F. Golden and D. Thompson, Cellular biology: Stem winder, *Time* 158, August 20, 2001.
8. K. Takahashi and S. Yamanaka, Induction of pluripotent stem cells from mouse embryonic and adult fibroblast cultures by defined factors, *Cell* 126, August 2006: 663–676.
9. J. L. Fox, Human iPSC and ESC translation potential debated, *Nature Biotechnology* 29, May 2011: 375–376.
10. R. Weiss, Scottish scientists clone adult sheep: Technique's use with humans is feared, *Washington Post*, February 24, 1997: 5.
11. Public Law No. 103-43, 107 Stat. 122 1993, NIH Revitalization Act of 1993. https://history.nih.gov/research/downloads/pl103-43.pdf.
12. J. Schwartz and A. Devroy, Clinton to ban U.S. funds for some embryo studies, *Washington Post* A15, December 3, 1994.
13. Public Law No. 104-99 128, Stat. 26, 34, 1996, The Balanced Budget Downpayment Act I.
14. H. B. Alberta et al., Assessing state stem cell programs in the United States: How has state funding affected publication trends? *Cell Stem Cell* 16, February 5, 2015: 115–118.
15. M. C. Nisbet et al., Framing science: The stem cell controversy in an age of press/politics, *Harvard International Journal of Press/Politics* 8, 2003: 36–70.
16. M. C. Nisbet, The competition for worldviews: Values, information, and public support for stem cell research, *International Journal of Public Opinion Research* 17, 2005: 90–112.
17. C. Fox, *Cell of Cells: The Global Race to Capture and Control the Stem Cell*, New York: Norton, 2007.
18. National Institutes of Health, *Estimates of Funding for Various Research, Condition, and Disease Categories (RCDC)*, February 2015. https://report.nih.gov/categorical_spending.aspx.
19. A. B. Parson. *The Proteus Effect: Stem Cells and Their Promise for Medicine,* Washington, DC: Joseph Henry Press, 2004.
20. M. Bellomo, *The Stem Cell Divide: The Facts, the Fiction, and the Fear Driving the Greatest Scientific, Political, and Religious Debate of Our Time*, New York: AMACOM, 2006; M. Ruse and C. A. Pynes, eds., *The Stem Cell Controversy: Debating the Issues*, Amherst, NY: Prometheus Books, 2006; C. Mummery et al., *Stem Cells: Scientific Facts and Fiction*, London: Elsevier Inc., 2011; L. Furcht and W. Hoffman, *The Stem Cell Dilemma: The Scientific Breakthroughs, Ethical Concerns, Political Tensions, and Hope Surrounding Stem Cell Research,* New York: Arcade Publishing, 2011.
21. Gallup, Stem cell research (the number of respondents who believe that research on human embryos is morally acceptable has stabilized at over 60%), http://www.gallup.com/poll/21676/stem-cell-research.aspx.
22. P. Knoepfler, Lessons from patients: Stem cell clinical trials unlikely options for most patients, The Niche, Knoepfler lab stem cell blog, January 14, 2013. https://www.ipscell.com/2013/01/lessons-from-patients-stem-cell-clinical-trials-unlikely-option-for-most-patients/; P. Knoepfler, How much do stem cell treatments really cost? February 22, 2015. https://www.ipscell.com/2015/02/stemcell-treatmentcost/; P. Knoepfler, *Stem Cells: An Insider's Guide*, Hackensack, NJ: World Scientific, 2013. Dr. Knoepfler is one of the leading stem cell research scientists in the Department of Cell Biology and Human Anatomy at U.C. Davis School of Medicine and also works for the Institute for Pediatric Regenerative Medicine.

2 Joining the conversation

What are stem cells? And why should you care?

Few advances in biomedical technology have captured the interest and imaginations of scientists and nonscientists alike as stem cells have. Human stem cell research is a young field—beginning in 1998, when scientists reported that they had isolated human embryonic stem cells. It is fair to say that few people had heard about stem cells or their potential for curing some of our most debilitating diseases when Dr. James Thomson from the University of Wisconsin and Dr. John Gearhart from Johns Hopkins University first published their findings. The coming decades would see the field erupt as scientists rushed to find a way to turn their research into clinical therapies. This chapter introduces the reader to the complex and multidisciplinary field of stem cell research. The first prerequisite is to reinforce your basic understanding of cell biology. Everything begins with the cell. Next, it is important to go behind the scenes and look at other medical and scientific discoveries that were essential to the development of the field. The development of in vitro fertilization, cloning technology, and skin grafts were all key factors contributing to stem cell advancements. Finally, after reading this chapter, you should have a better idea of what is currently possible and what is not. Distinguishing between hype and hope is not always easy. Some therapies have been around for decades (such as bone marrow transplants), but labeling them as new or improved may be just an excuse to increase patient fees. After reading this chapter, you should be able to tell the salesman from the scientist.

Cell biology in a nutshell

You might recall from high school biology that the human body is made up of trillions of cells. Each cell is a living unit with a complex structure; most cells in the human body are designed to carry out specific functions. These cells, which we will refer to as adult cells (although scientific researchers prefer to call them somatic or tissue cells), have the ability to self-renew and to turn into specialized cells that can replace or repair other cells in the tissue.[1] Skin and blood cells are among the most prolific at dividing into copies of themselves. Individual cells are short-lived, some lasting only a few days or a few weeks. Cell renewal rates vary depending on the tissues in which they are found—we are continually losing our hair, yet most of us never go bald. It is the stem cells within the various tissues of the body that produce all the cells the body needs to stay alive.

Stem cells, found throughout the body, have the ability to divide (duplicate); they also can produce all the specialized cells within the tissue in which they are located. It should be noted that while the term stem cell dates back to the mid-nineteenth century, it was not until scientists began to study blood that they developed a better understanding of the existence and function of stem cells.

A stem cell can give rise to two types of cells. One of these will become a specialized (scientists prefer to use the term differentiated), mature cell; the other will remain a stem cell and retain the capacity to divide again, replicating itself as well as making the specialized adult cells needed for a particular tissue. This is called asymmetric division. Instead of giving rise to two of the same cells, this division yields an adult cell and another stem cell, which allows the cycle of replication to continue as long as is necessary. These cells are referred to as multipotent because they can turn into any of the cells needed for a specific tissue. They cannot, however, form all the cells needed to create an entire person. For this, we need embryonic stem cells.[2]

Human embryonic stem cells, first isolated in 1998, have the unique characteristic of not only dividing indefinitely (under the right conditions), but also of differentiating into all the cell types normally found in the human body. Embryonic cells come from cells that lie within the early (4- or 5-day-old) embryo; a single cell is removed and put on a layer of culture medium where it will continue to grow under the right conditions. These cells are labeled pluripotent because they can and do (when implanted in the uterus) become an individual. In this chapter we will be discussing embryonic stem cells that exist only in the world of in vitro tissue cultures.[3]

To review, stem cells are cells with the ability to divide (indefinitely) so that they can renew themselves, and they can give rise to differentiated cells. Adult stem cells are considered multipotent, while embryonic stem cells are considered pluripotent.[4]

Why should you care about stem cell research?

Technology has helped many people who need to have a body part replaced as the result of disease or accident, and it continues to do so. Some parts of the body, such as joints and teeth, are replaced with man-made materials. Others need human replacement—the heart, the liver, and the cornea are transplanted from people who have died. Still other parts, such as blood, bone marrow, and kidneys, can be taken from living donors. In some cases, a person's own body can be used to supply replacement parts, as with skin grafts. The number of people waiting for a human transplant far exceeds the number of available donors. Sometimes the recipient's body will reject the transplanted organ; this is referred to as graft-versus-host disease and is a common side effect of bone marrow transplants.[5] What if you could circumvent these issues by replacing or repairing diseased or aging organs with stem cell therapies? This is the hope of stem cell research.

Today, stem cell research has the potential (and we stress potential) to cure our most debilitating illnesses.[6] The field of regenerative medicine is turning some early developments into human clinical trials. While you may be a healthy individual, it is likely that you know someone who is suffering from a disease such as cancer, Parkinson's, or Alzheimer's, or from a spinal cord injury. Learning what stem cell research can do when applied is reason enough to stay informed about this expanding field.

How much are you willing to pay for these therapies? The terms personalized medicine, precision medicine, and individual therapy are being used more frequently. Early and hypothetical articles suggest that costs for some treatments may exceed half a million dollars.[7] If your health insurance company is not willing to pay for expensive new treatments, should society pay? Should the federal government pick up the tab for new treatments if they are proven to extend life? This is not an easy question to answer; we might need to weigh the options. Will we expect our health insurer to cover some of these experimental and/or expensive therapies? This is an equity issue that we will tackle in Chapter 6. Who benefits and who pays are the thorny questions underlying all matters of public policy and, especially, medical breakthroughs.[8]

Cells working together for a healthy body

It is probably fair to say that most of you have forgotten much of what you learned in high school or college biology. If you fit into this category, the next section will refresh your memory or send you to one of many basic biology texts.[9]

The human body is made up of trillions of cells. Inside each cell are even smaller structures called organelles that control, produce, and move materials; release energy; and work together to keep the cell alive. Although cells come in different shapes and sizes, they all share the same basic structure. Each cell has a membrane, or an outer layer that surrounds the cell. Inside the membrane is a liquid called cytoplasm that supports the different organelles. The nucleus contains a long complex molecule called deoxyribonucleic acid (DNA). You have seen that acronym many times.

Figure 2.1 illustrates a basic cell adapted from the Human Genome Project and illustrates the genetic information (DNA) found in the nucleus of each cell that is unique to you; the DNA also tells each cell what kind of cell to become and what to do. The instructions, called genes, come in the form of chemical codes in the DNA. As part of 46 coiled lengths known as chromosomes, some genes work while others are switched off, which is why cells are different and perform varied tasks. When a cell divides, it copies its genes and then passes them along to its offspring cell. For more information about this process, see the Human Genome Project.[10]

While each cell in your body contains the same DNA, the instructions within each cell will vary depending on where in the body (in which organ or tissue) the cell is located. Remember, blood cells can replicate themselves but cannot make skin cells.

DNA looks like a twisted ladder. The cells in your body all have the same DNA. Half came from your mother, and half came from your father. Unless you have an identical twin, no one else has your exact DNA. This is why on-screen (and occasionally real-life) detectives can use

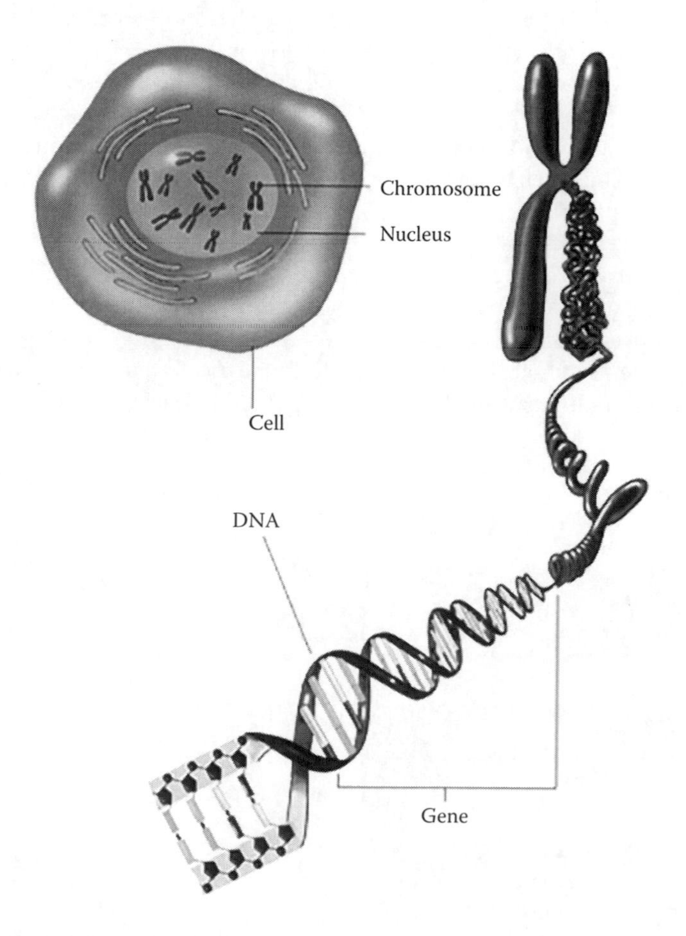

Figure 2.1 Cell and DNA (Adapted from National Institutes of Health, National Human Genome Research Institute, *Fact Sheets on Science, Research, Ethics and the Institute*. https://www.genome.gov/10000202.)

DNA to determine if a person committed a crime. Your bodily fluids and even hair or skin samples contain your unique DNA.[11]

Human cells have different shapes and sizes depending on where they are found in the body. Some are long and thin, some round and fat. Others, like nerve cells, have a spidery appearance. Figure 2.2 is a depiction from the National Institutes of Health (NIH) of what some of the cells in your body look like. If you are interested in viewing additional realistic depictions of this nature, you might go to the NIH website or check out a text on human cell biology.[12] The human body contains roughly 200 kinds of cells that carry out different functions. So, while each cell in your body contains the same DNA, the instructions within each cell will vary depending on where in the body the cell is found. Cells do not stand alone; they hang out together with other similar cells.

Like all living things, cells wear out and die. Cells can also be destroyed when a body is injured or grows sick. Dead skin cells flake off and add to household dust, and hair cells come off and clog your sink. Dead cells from inside the body are broken up and eliminated with waste. Some dead cells stay in place. For example, our nails are made up of hard layers of dead cells that protect the cells underneath from damage.

Cells that do the same job are grouped into the many tissues that are found inside the body. There are four main tissue types: epithelial, connective, nervous, and muscle. The following information is synthesized from the National Institutes of Health website discussing the four types of cell tissues.[13]

Epithelial tissue covers the outside and lines the inside of most of your body parts. Cells in epithelial tissue are tightly joined to one another. Together they form tough, stretchy sheets. Epithelial tissue even covers the eyeballs and the inside of the nose. One of the clinical experiments currently being conducted is trying to cure macular degeneration by growing new eye cells from human embryonic stem cells (hESC).[14] Your skin is the most obvious epithelial tissue: Sheets of epithelial cells form the skin's outer layer. New epithelial cells are always being formed beneath the skin's surface and these cells move up to replace dead cells. All your skin's epithelial cells are replaced every 75 days. Much of the dust in the average home contains a lot of dead/discarded epithelial skin cells. Some of the first experiments growing skin for burn patients took place in a lab.[15] The scientists used adult skin cells from the burn patient to grow new cells in a lab, put them on a backing, and placed them on the patient's skin. The lab-grown tissue eventually covered the burned area.

Connective tissue is the most abundant tissue found in the human body. It holds things together and comes in different forms. Beneath the epithelial tissue in your skin is a layer of

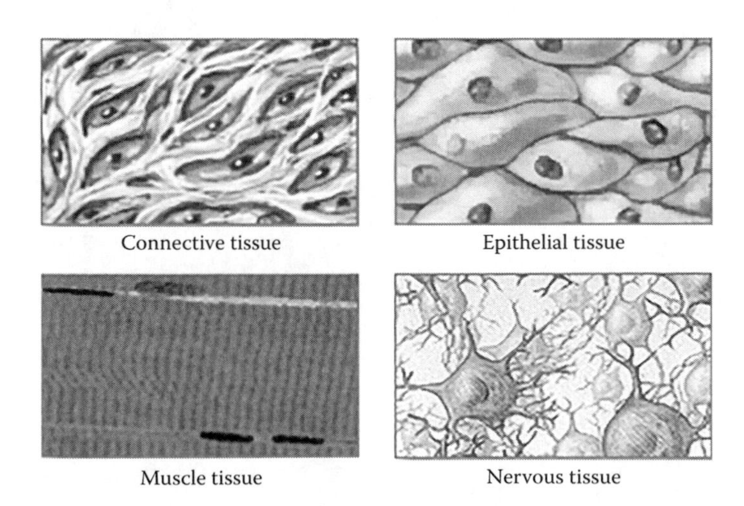

Figure 2.2 Four Types of Tissues (Reproduced from NIH, National Library of Medicine.)

loose connective tissue. Fat tissue is a special type of loose connective tissue that cushions you from bumps and falls. Fat tissue also help keep you warm. Liposuction is a common procedure used by plastic surgeons to remove unwanted fat from one part of the body and, sometimes, reinsert the fat in another part of the body.

Along with fat, tendons are dense connective tissue made up of strong, stretchy fibers. These fibers all run in the same direction and connect muscles to bones. Bone is another connective tissue consisting of several different tissues working together—bone, cartilage, dense connective tissue, epithelium, blood-forming tissues, adipose tissue, and nervous tissue. Each individual bone is an organ. Wherever you find bone, you usually find cartilage, which is also a connective tissue. It is made of cartilage cells and fibers that form a rubbery matrix. Blood is also a connective tissue, albeit a liquid one. Blood contains white blood cells, red blood cells, and platelets. Platelets are tiny cell pieces. All your blood cells float in a watery liquid called plasma. Blood has two main jobs: It brings oxygen and nutrients to cells, and it carries away their wastes. In this way, blood connects all the parts of your body.

Nervous tissue is the third major kind of tissue and is composed of neurons. Neurons' cells enable them to receive and facilitate nerve impulses. This tissue is the main component of the brain and the spinal cord. Nervous tissue has lots of nerve cells, but it also has even smaller supporting cells that hold nerve cells in place. Neurons are classed in several different ways. For our purposes it is enough to remember that nerve cells send and receive messages.

The fourth type of tissue is *muscle tissue*. In muscle tissue many muscle cells contract at the same time. Not all muscle tissue in your body is the same. Skeletal muscle tissue is made of long, sturdy cells that contract very quickly. This tissue has enough power to move whole body parts. Smooth muscle tissue is made from shorter, thinner muscle cells that contract more slowly. Smooth muscle helps move things through your body, like food through the stomach. Heart muscle tissue is only found in your heart. Heart muscle cells are short, branched, and tightly connected. This close connection helps heart muscle cells work well together. Heart muscle tissue is the superstar of muscle tissues. It never gets tired of contracting, and, during your lifetime, it will never stop. But your heart is made up of more than just heart muscle tissue. It is also home to epithelial tissue, connective tissue, and nervous tissue. Your heart is a combination of all four different kinds of tissue working together.

The four main tissues in your body are organized to form organs. Tissue groups work together within organs to keep your body going.

Humans have around 22,000 genes. Some genes are dominant, and they beat out genes that are recessive. The completion of the Human Genome Project in 2003 discovered how genes communicate.[16] It was once thought that one gene controlled a particular body/cell feature; however, recent discoveries about the body show that many features are controlled by several genes working together rather than just one. We don't want to turn you into a geneticist, but it is important to understand that a cell gets its instruction from its DNA. This concept was/is a key factor in the discovery of induced pluripotent stem cells (iPSC).[17] By figuring out what genes are needed to turn an adult/specialized cell back into an embryonic stem cell, the researchers used chemicals to transcribe the genes. A length of DNA has a double helix shape and resembles a long twisted ladder. This ladder's rungs are made up of four chemicals called bases—adenine, cytosine, guanine, and thymine. These bases are always linked in pairs: adenine with thymine, and cytosine with guanine. A specific order of bases forms an instruction, called a gene, that controls a part of the body, such as eye color or skin color.

Stem cells move to center stage

The term stem cell has been around since the mid-1800s, but back then the term did not receive the attention that it does today. Outside of the scientific community, few politicians, reporters, patient advocates, or patients thought much about the topic. In 1998, two scientists published research showing that they had grown embryonic stem cell lines in vitro and things began to change.[18] Before we get to the work of James Thomson and John Gearhart, let's take a detour to put this research into scientific perspective.

In order to set the stage for the discovery of embryonic stem cells, we need to review a few events that led up to this discovery.

First, while only a few names are associated with the discovery of human embryonic stem cells (hESC), hundreds of scientists all over the world were working on related research.[19] Further, scientific research is done in labs with dozens of people working on different aspects of the same experiment. It is enough just to look at all the names listed on a published scientific paper to realize that one person did not do it alone. When we move into the field of regenerative medicine, larger multidisciplinary teams of researchers are needed that might include engineers, physicists, and social workers. The personal computer may have been developed in a garage by a group of high school students, but, rest assured, stem cells were not grown in someone's kitchen.

It is also important to remember that much of the research that is tested on humans is first tested on animals (usually mice, but also larger animals). Scientific discoveries must be replicated and reproducible. Making sure that the research is accurate can take a long time, which makes for a long road from Petri dish to drugstore.

Finally, stem cell research owes a huge debt to work done on bone marrow, skin transplants, in vitro fertilization (IVF), and animal husbandry. We will discuss these contributions here to help you see a broader picture of this research.

Bone marrow transplants and adult stem cells

Although scientists knew a great deal about human blood and had hypothesized that it contained different types of cells, it was experience with radiation sickness that led to the discovery of adult stem cells. Doctors treating the survivors of the atomic bombings of Hiroshima and Nagasaki could not understand why individuals with no outward signs of injury began to die within a few weeks. It was eventually discovered that they were suffering from radiation sickness.[20] The radiation from the bomb destroyed the stem cells in their blood; without a continuous production of red blood cells, white blood cells, and platelets, the body began to shut down. Too much radiation can have a deadly effect on the human body. When you get an x-ray today, the technician will put a lead apron over the part of the body not being targeted by the radiation, and he/she will step out of the room. This is to ensure that you do not receive too much radiation, which can add up over time and may lead to cell mutations. Maybe you know someone who has been treated for cancer. If the treatment included radiation (to kill all the cancerous cells), the radiologist calculated how much radiation could be administered to the individual. Radiation kills cancerous cells, but it can also kill healthy cells. Blood diseases such as myelodysplasic syndrome, the inability of the bone marrow to produce enough white blood cells to fight off infections, have been linked to excessive radiation.[21]

Blood is made up of trillions of cells floating in a watery liquid called plasma. There are three main types of blood cells. Red cells transport oxygen from the lungs to the tissues; white blood cells kill disease-causing germs; and platelets assist in creating blood clots that help heal wounds.

The existence of adult blood stem cells was confirmed in the 1960s with experiments by two Canadian scientists, James Till and Ernest McCulloch.[22] They proved that a single cell from a person's bone marrow could generate copies of itself and multiple types of blood cells (red, white, and platelets). They reached this conclusion by irradiating mice (mimicking what had happened to the citizens of Hiroshima and Nagasaki) and then injecting the irradiated mice with bone marrow from healthy mice to see how many of the healthy blood cells were needed to restore the damaged blood cells. In subsequent experiments they tracked the cells that they injected (typically done using a dye) and discovered that a single cell from the bone marrow of the healthy mouse could generate copies of itself and the multiple types of blood cells needed to keep the sick mouse alive.

A subsequent experiment uncovered something else about stem cells: They were also capable of existing in a state of arrest, which suggested that adult stem cells that exist in the tissue of organs possibly last the entire life of the animal.[23]

With the bloodstream hoisted high as a model, it suddenly seemed highly likely that other tissues must also harbor rare populations of stem cells.

In tissues like blood, skin, or intestines that turn over frequently, it was reasonable to hypothesize that there must be a lot of stem cells, but in tissue that does not turn over, like neurons in the brain, isolating them to identify stem cells is not so easy and remains a problem today. It can be done on lab animals by using dyes to track the injection of the cells, but this is not something that can be done on humans for routine medical trials. We know whether someone had Alzheimer's by performing a postmortem autopsy, but that may not be very helpful in trying to find a cure for a person living with the disease.

The first bone marrow transplant was completed in 1956 on two identical twins. The key here was that since both donor and host were a perfect genetic match, there was no rejection. Subsequent transplants were done first on siblings and later on unrelated individuals. Today, bone marrow transplants are a common procedure for individuals suffering from blood-related diseases such as leukemia. However, the transplants are not always successful because patients need a close genetic match lest the body reject the transplant. Improvements in immunosuppressant drugs help some patients. Today, more efficient methods for harvesting hematopoietic stem cells have become more common. While stem cells are still drawn from bone marrow, more frequently they are harvested directly from a patient's bloodstream (the peripheral blood). A person is given an injection that stimulates the marrow to produce greater than normal amounts of stem cells, that in turn spill into the bloodstream. The individual's blood is then drawn, from which the stem cells are gathered and put aside for transplantation. In 2013, an estimated 19,000 transplants and 15,000 allogeneic transplants were performed with peripheral blood stem cell grafts in the United States, according to the International Bone Marrow Transplant Registry.[24]

Umbilical cord blood is another source of stem cells that can be used to treat individuals (usually children) with blood disorders. Despite the controversy regarding the utility of the practice, more parents are choosing to collect and store the blood from their newborn's umbilical cord.[25] A number of private cord banks arrange for collection and storage of this blood for eventual use by the child (or someone else in need of a transplant). Public cord blood banks also exist, to which parents can donate their newborn's cord blood for use by anyone who needs a transplant. The problem is that the amount of blood taken from the typical umbilical cord is very small (on average 5–12 tablespoons), and while it may be enough to restore a child's blood system, it is usually not enough to restore an adult's hematopoietic system.[26] The use of cord blood also plays a role in another controversial topic—the "savior sibling"—that will be discussed later in this chapter.

Therapy with cultured cells: Skin grafts and beyond

While blood connects all of the major organs in the body, the skin is the largest organ. It consists of two main layers that are basically glued together. The outer layer, or epidermis, lies on and is nourished by the thicker dermis, which contains the blood vessels, nerves, sweat glands, hair, and oil follicles. The surface of the skin (think palms, fingers, and feet) is constantly being worn away. So it was reasonable for scientists to hypothesize that there must be some unspecialized cells that divide to form the top layer of replaceable cells, while others (stem cells) remain in place to divide again. Is this starting to sound familiar?

About the same time that scientists were researching blood stem cells, Howard Green and his colleague George Todaro were trying to figure out how to grow cells in vitro.[27] Cell lines were not particularly new. In the 1950s, George Gey had grown cervical cancer cells from Henrietta Lacks and turned them into the HeLa cell line that is still used today.[28] But the problem with these cancerous cells and subsequent noncancerous cell lines is that they had the tendency to cause tumors when injected in vivo. So Green and Todaro, his young medical student, set out to systematically create a culture that could be used to grow normal cells. They wanted an experiment that could be replicated by others. (Note that this is one of the key tenants of scientific discoveries: It is not enough to say you did it; it is essential that other researchers can attain the same results following your protocol.)

In 1962, Green and Todaro used mouse embryonic fibroblast cells (from Albino Swiss mice) that were cultured using a strict protocol they established. The colonies were grown on beds of connective tissue cells called fibroblasts; the scientific literature indicates that such feeder cells often help other cells proliferate. After growing the cells in vitro for about 20–30 generations, they found that the cells began to stabilize. Today, these mouse fibroblasts are still used by researchers as a bed for growing cells. This is why Green is sometimes called the father of stem cell research.[29]

Having developed a stable medium on which to grow cells, Green and another colleague, James Rheinwald, discovered cells resembling epithelial (skin) tissues in the tumors or teratomas of mice.[30] Scientists typically use mouse models to test their early theories. At the time, epithelial cells could not be cultured in laboratories. The researchers isolated the epithelial tissues from the tumor and added the 3T3 (feeder) cells, and they soon found colonies formed with structures resembling the outer layer of skin. The team had succeeded in replicating the basic building blocks of the epidermis. Green's experiment was the breakthrough needed to show that cells can be grown in vitro and then transplanted into humans.

The discovery was fortuitous, for in 1983 Green was contacted by doctors from Shriners Hospital for Children—Boston seeking help for two brothers from Wyoming who were being flown to the hospital with burns covering 90% of their bodies. Green and his team took skin samples from the boys' armpits, grew the small samples in culture, and then turned their lab into an around-the-clock skin factory. Remember the 3 in the 3T3 formula? It signifies that every 3 days the cell tissue you are growing has to be replated (moved to another dish with fresh-growing medium); otherwise the cells might die or grow into clumps and be rendered useless. Maintaining tissue cultures in vitro requires precision and timing. Waiting too long could mean that the cells have died or have started to specialize in ways that make them unsuitable for human transplantation. It took about 14 weeks for Green's team to grow enough skin to cover the boys' burned bodies. Both boys had to undergo more than 100 skin grafts and, although they survived, their quality of life was diminished. Both boys suffered third-degree burns, which meant that their sweat glands were also destroyed; Green could grow the top layer of skin but not the dermis. Today, several companies still use the method developed by Dr. Green to grow skin for burn victims.[31] Major burns have decreased substantially, but there are still a few companies that grow skin for burn patients based on Green's original process.

Howard Green has been called the founding father of regenerative medicine. He accomplished what all researchers hope to do—save people's lives with their experiments. Today, skin grafts are commonly grown in culture and then transplanted onto the burn patient. One of the first approved hESC experiments also used a technique developed by Green and his colleagues to restore vision to individuals with macular degeneration using corneal stem cells.[32]

IVF and embryonic stem cells: It all started with Louise Brown

If you have never heard of Louise Brown, you are probably not alone. In 1978, as the first baby in the world born using a new experimental reproductive technology—in vitro fertilization (IVF)—hers was a household name. In short order, her birth was followed by that of the first IVF baby born in Australia, and finally, in 1981, the first IVF child was born in the United

States.[33] In 2013, a party was held in England to celebrate Louise's 35th birthday. At the party was Louise's mother, her fertility doctor, her husband, and her son, who had been conceived naturally. All this to prove that she led a normal life, thus disproving some of the early concerns that so-called Petri dish babies might be abnormal and suffer from physical and/or psychological conditions. Louise Brown published a book describing her life in 2015, in which she notes that her parents were criticized initially for engaging in this new therapy.[34]

So what does the therapy involve? The procedure, in theory, is very simple: Eggs are removed from a woman's ovary and placed in a Petri dish, where they are fertilized by sperm from a donor. Once the eggs are fertilized, embryos are created. After 4 or 5 days, a few embryos are implanted into the womb of the woman and, if everything goes well, a child is born in 9 months. The theoretical simplicity obscures the many unexpected problems that can, in practice, occur. Perhaps the egg is not fertilized; the egg is fertilized but stops growing before it is implanted; the egg is implanted but is miscarried before it matures; or the baby is born but is affected by a life-threatening illness. Theory and practice do not always align to achieve a happy ending. In fact, it is estimated that only about 25–30% of all IVF treatments are successful at producing a live birth.[35] The treatment can also be very expensive.

By now you have figured out that creating embryos in a dish is not a perfect science.

Sizable sections of the religious world thought that IVF was equivalent to scientists playing God. Some predicted that the children would be born with abnormalities, some foresaw that this could lead to surrogacy issues, and others proclaimed it a blessing for infertile couples. To this day, the Catholic Church prohibits any form of reproductive technology, believing that only natural procreation is allowed.[36]

Irrespective of where you stand on the morality of this issue, the field of reproductive technology is a booming business. We stress business because no federal money was ever spent on IVF research. This means that IVF clinics in the United States avoided federal supervision or regulation, and it also meant that each clinic could do as it pleased.

Today, there are over 200 IVF clinics in the United States. There are several membership groups that exist to try to bring some consistency to the field. One such group is the Society for Assisted Reproductive Technology; another is the American Society for Reproductive Medicine.[37] However, the fact remains that there are no government regulations regarding what fertility clinics can charge for their services. Today, it is not uncommon for unmarried women to use IVF with donor sperm to have a child. This was not permitted in the early days of IVF. Dr. Jones notes in his book that the donor couples (that is, married couples) were carefully screened. Initially, only women with blocked fallopian tubes were eligible for IVF treatments.[38]

Today, some employers (usually in the private sector) provide IVF coverage in health care plans, although most do not due to the expense. Recently, a couple of Silicon Valley companies made news by paying interested female employees to have their eggs extracted and frozen for future use.[39] You may have also read about couples going to court to determine who owns an unused/frozen embryo.[40]

What began as an exciting new option for the infertile couple opened up the proverbial Pandora's box of issues both personal and public. For example, when does life begin? Who owns the unused embryos? Is surrogate motherhood legal? These questions and others will be explored in greater depth in Chapter 6. The short and unsatisfying answer is: It all depends.

More to the point of our research is, what happens to all the unused embryos derived for IVF treatments? Scientists had been studying mouse embryos for more than 20 years,[41] so they were eager to put their knowledge to use on human embryos. But getting human embryos (or being allowed to use human embryos) depended on where you were doing your research. Let us stick with the United States, where there was a lot of confusion between state and federal governments and between public and private funds.

In the United States, support for IVF was controversial. As early as 1971, the Human Embryo Research Panel thought that, under certain conditions, research involving fetal and embryonic tissue should be supported.[42] But the panel's report was ignored. In 1973, the U.S. Supreme Court legalized abortion in *Roe v. Wade*. According to Howard Jones, this decision seemed to also ensnare IVF research, although IVF research was never federally funded. The first clinic to offer treatment in the United States was founded in Norfolk, Virginia, at Eastern Virginia Medical School.[43] So while IVF clinics are technically private institutions, they are caught in the middle of the battle over what to do with unused embryos.

IVF created another unintended consequence—that of preimplantation genetic diagnosis (PGD). We do not hear too much about this anymore. A few sensationalized cases occurred around 1990, and perhaps this procedure is still controversial but is now flying under the radar. Or maybe no one cares anymore.

Once you can create an embryo in a dish, you can then proceed to check the cells for potential genetic diseases. We will use the example of Molly Nash for illustrative purposes, but know that there were several other cases that created quite a stir. Molly Nash (a real person) was born with the genetic disorder Fanconi anemia and was not expected to live past the age of eight or nine. She was in need of blood transfusions and was susceptible to immune disorders. Undergoing a bone marrow transplant could ensure that her blood stem cells could regenerate. You are now familiar with bone marrow transplants, but remember the key is that results are most successful when there is a tissue match between the donor and the host. So, Mr. and Mrs. Nash decided to have another child using IVF so that embryos could be tested and those that were tissue matches with Molly would be implanted. You also know now that IVF does not always work, so the chances of success were slim. But the Nash family went ahead, and IVF produced a viable embryo with a matching tissue type. The child was born and the blood from his umbilical cord was retained and injected into Molly. The procedure worked. Both Molly and her brother, Adam, are alive and well.[44]

But the controversy did not end quietly. Some felt that it was immoral to have one child to save another. Children should not be used for instrumental purposes. The Catholic Church condemned the Nashes behavior. The Nashes, for their part, pointed out that they were Jewish. In England, there was a national debate that revolved around two similar cases that fractured the medical community and pitted doctors on both sides of the issue against one another.[45] The 2004 book *My Sister's Keeper*, which was later adapted into a film, was a thinly veiled version of the Nash story with an unhappy ending.[46]

We suspect that this type of procedure may still be going on without generating as much controversy today because, as we noted, IVF clinics are private; they need not answer to regulatory bodies (at least not in the United States). As we have maintained throughout this chapter, novel or breakthrough procedures make headlines, but when they become routine, they lose their shock value and news appeal.

IVF has opened up a host of issues that are not the topic of our book: surrogate mothers, nonbiological children, embryo adoption, nonbiological embryo implantation, and even designer children. If you can pick an egg and a sperm from high-functioning donors, why not do it? Or is this the next step to the designer babies that some ethicists are concerned about?[47] This is a topic for another book.

Since IVF was developed without federal/public funding, it also avoided federal supervision or regulation—and whether or not that is a good thing is up for debate. Clinics could decide who to treat, what to charge, and what to do with the unused embryos. Most embryos remain frozen for years; some, however, are donated to private labs for research—the kind of research that eventually led to the creation of human embryonic stem cell lines.

Before we move to the discovery of human embryonic stem cell lines, there is one other topic that must be explored—cloning. Scientists have long been interested in fertility. Mostly, however, it was animal and plant fertility. Artificial insemination was and still is regularly used in the selective breeding of animals. So how does a cloned sheep named Dolly fit into the picture?

A sheep named Dolly: Cloning animals

In the years between improving on the success rate of IVF and growing animal embryonic stem cell lines, a number of scientists were working in another area that became controversial—cloning animals.

A clone is a copy. How many times have you wanted to clone yourself so that you could be in two places at once? While the expression is used in common speech to reflect the stress of a fast-paced society, in the biological world it has a more specific and achievable result. In works of fiction, cloning is portrayed as a real outcome that both entertains and upsets. In the book by Ira Levin, *The Boys from Brazil*, the evil Dr. Mengele ends up cloning Hitler many times in the hopes of recreating the Third Reich—great story, but dubious science.[48]

While we use the word clone to refer to an exact copy, the more scientific term used for the biological procedure is somatic cell nuclear transfer (SCNT). Sometimes changing the name of a procedure can also change the cognitive connection and, ultimately, the approval or disapproval of it. What you call something matters in the world of science as much as it does in other fields. While some of this is about political correctness, it is also about scientific precision—making sure that the procedure can be replicated and the results will be the same.

We should point out that molecular cloning is done in most biomedical laboratories.[49] You can clone cells in tissue culture. You might remember we discussed the growing of skin cells in which the sections of skin cells were all grown from one cell. The connection between cloning and embryonic stem cells will become apparent very soon. By combining two procedures, Shinya Yamanaka was able to create induced pluripotent stem cells (iPSC).[50] But first, back to Dolly.

Dolly the sheep was the first mammal to be cloned. In the late 1950s, John Gurdon from Oxford University was the first to perform a similar procedure with frogs. He was able to generate cloned frogs from tadpole cells. When he used the cells from adult frogs, tadpoles developed that were identical genetically to the frogs from which the donor cells were taken.[51]

The theory behind SCNT is easy to understand (See Figure 2.3).[52] An unfertilized egg cell (oocyte) is removed from the donor animal (it could be from any sheep, but in Dolly's case it was from a Scottish Blackface). A mammary cell was taken from another sheep (in Dolly's case, from a Finn-Dorset). The nucleus in the unfertilized egg was replaced with the nucleus of the mammary cell. The two were fused together using an electrical charge and the fused cell was placed in a Petri dish and grew into a blastocyst (a mass of about a dozen cells that, when implanted into the uterus of another sheep, would produce an offspring). It may sound easy, but it is a little more complicated than that. To begin with (as with IVF), the success rate after implanting the embryos is very low. With Dolly, more than 277 unfertilized eggs were used before scientists at the Roslin Institute in Edinburgh, Scotland, produced a live birth.[53] Not a very good success rate if you are thinking of doing this on a commercial basis.

Dolly became famous worldwide, but not everyone was happy about the experiment.[54] Dolly was born in 1996 and was euthanized in 2003 after she developed arthritis and lung disease. She lived a charmed life at the Roslin Institute, and she is now on display at the National Museum of Scotland.

After Dolly, researchers around the world began to clone other animals. In the United States, Dr. Jerry Yang, a researcher from the University of Connecticut, began to clone cows that would produce more milk.[55] Yang worked with research teams in China and is credited with jump-starting China's stem cell research efforts at a time when researchers in the United States were limited in the human embryonic stem cell lines available.

After Dolly, the most famous cloned animal was an Afghan dog named Snuppy (for Seoul National University Puppy), cloned in 2005, after many years of research, by South Korean scientists. Snuppy was named as *Time* magazine's Most Amazing Invention of the year in

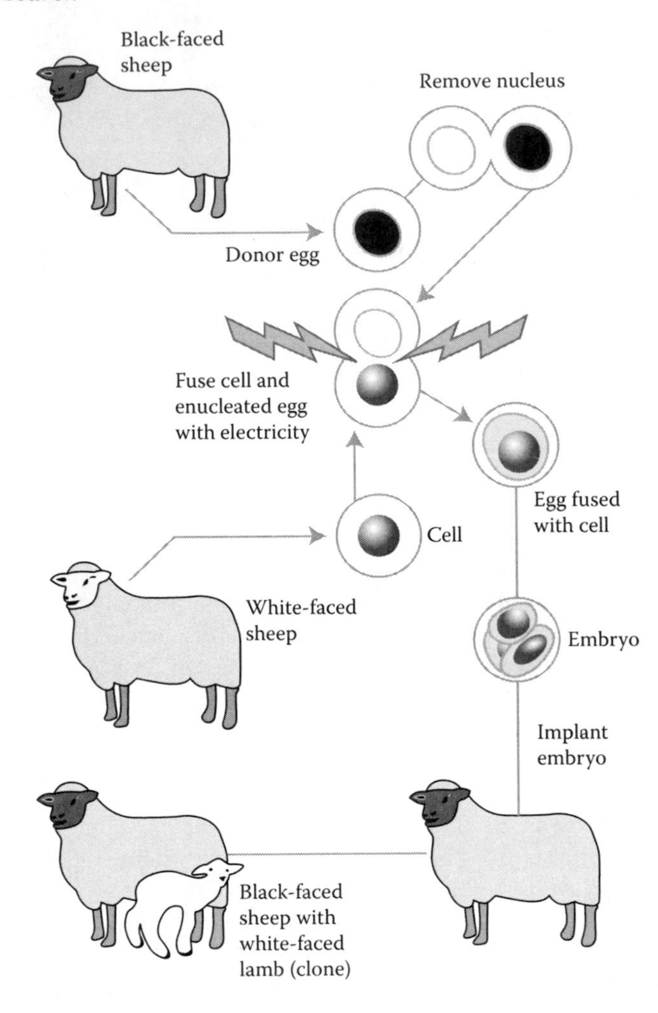

Figure 2.3 Cloning Dolly (Reproduced from *How Stuff Works*.)

2005.[56] Not everyone was pleased with the results, though. Even the South Korean researchers noted that the low success rate made the effort interesting but of little practical value. In short, it took a lot of unfertilized eggs to create a clone.

The commercial value of cloning animals with specific sought-after traits did gain some notoriety. Commercial services advertise on the Internet. One particularly short-lived company called Genetic Savings and Clone was located in Sausalito, California, from 2004 to 2006.[57] The company was financially supported by John Sperling, the billionaire and founder of the University of Phoenix. It advertised that it would clone your pet for $50,000, but there did not seem to be a big market for cloned pets. Commercial cloning of farm and sports animals, however, is a more enterprising endeavor. For example, in 2013, Show Me, a six-year-old clone of a polo-playing mare named Sage, won the Argentinean Triple Crown.[58] It is interesting to speculate how sports involving animals would change if this procedure becomes more common.

The successful cloning of Dolly proved that: (a) many other large mammals could be cloned, and (b) an adult cell, when placed in the nucleus of an unfertilized egg, could turn back and become a pluripotent stem cell that would be genetically matched to the patient. It took more than a decade before scientists in Japan improved on this idea to discover induced pluripotent stem cells.

U.S. researchers grab the human embryonic brass ring (hESC)

In 1999, *Science* magazine voted human embryonic stem cell research its breakthrough of the year.[59] While it is safe to say that the general public did not know much about this scientific advancement, within a few years it would be the topic of much political controversy. In fact, it is still a controversial topic.

In 1998, a team of researchers under James Thomson at the University of Wisconsin–Madison derived hESCs from spare embryos created in IVF clinics. Thomson's team cultured the cell lines for months without differentiation (specialization) and then induced the lines to differentiate into the main groups of embryonic tissue layers (ectoderm, mesoderm, and endoderm).[60] The team working under John Gearhart at Johns Hopkins University successfully isolated and cultured human embryonic germ cells, derived from aborted fetuses.

It is interesting to note that while Gearhart directed the IVF lab at Hopkins during the 1980s, he decided, after consultation with university administrators, that working with human embryos might lead to ethical controversy. By using tissues from aborted fetuses, his team could achieve the same goal—growing embryonic germ cells. Germ cells, which come from the genital ridge of the fetus (cells that represent the premature testes or ovary), when removed and cultured, give rise to every type of cell in the human body.

Both Gearhart and Thomson owed their successes to the work done decades earlier by Drs. Gail Martin and Martin Evans.[61] From mice, researchers moved on to monkeys and other primates. But prying cells from human embryos required more time and skill. You cannot just stick the cell in a Petri dish and forget about it. The keys to cell growth are precise timing and feeder cells that keep the cell continuously dividing while limiting its natural inclination to differentiate. Figure 2.4 illustrates the steps in creating an hESC.

Gearhart and Thomson also owed their successes to another individual, biomedical scientist and wealthy entrepreneur Michael West. In 1990, he founded the Geron Corporation to research the aging process. Over the years, he has been associated with several other biomedical

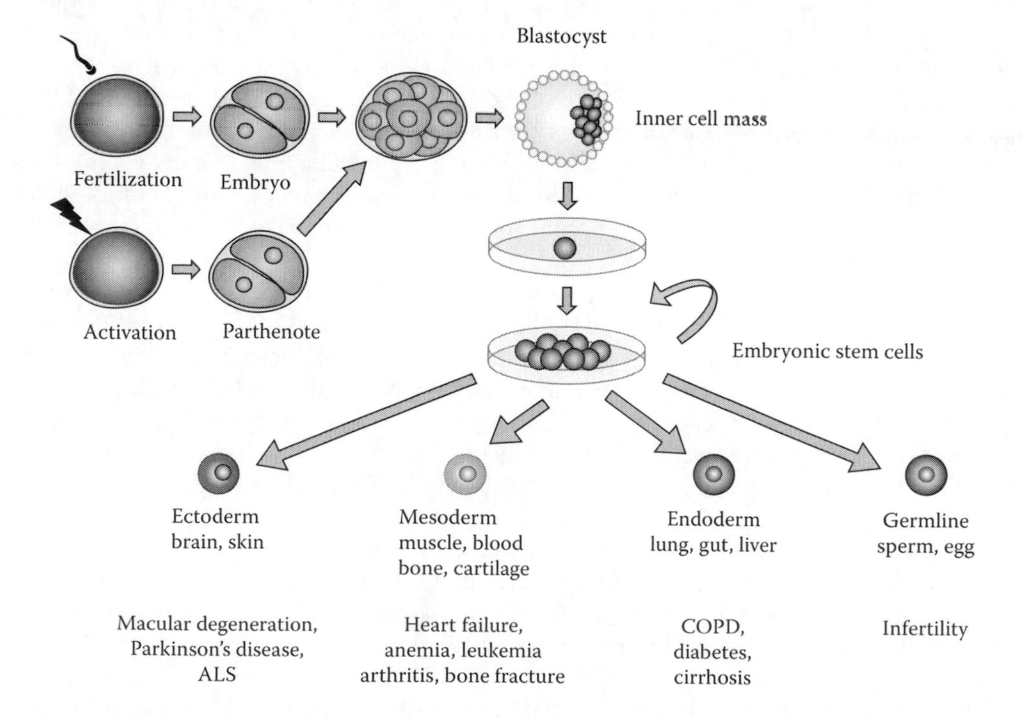

Figure 2.4 Deriving Human Embryonic Stem Cells (Reproduced from pinterest.com. https://flowvella.com/s/ozj/C8568623-16C0-4867-BADB-301B5F6E34E2.)

technology companies. West funded the research of both Gearheart and Thomson.[62] Access to private funds allowed both researchers to grow their cell lines without public funding.[63] In exchange, the researchers shared patent rights with West. The Wisconsin Alumni Research Foundation (WARF) held the patents on the stem cell lines derived by Thomson. The patents initially created a problem for scientists who wanted to work with the cells; researchers had to pay WARF for the cells and initially had to agree to report back to WARF annually on the status of their research. Since a private company had funded the research, WARF was compelled to honor a licensing agreement with Geron.[64] Once researchers figured out other ways to grow their own stem cell lines, the field began to grow.

Expanding the stem cell commons

Stem cell research is an international field. In other parts of the world, researchers were already working with human embryos. In her book *Cell of Cells: The Global Race to Capture and Control the Stem Cell*, science journalist Cynthia Fox describes the work going on in various countries to establish world-class research teams.[65] The subtitle of the book says it all: Countries were and are spending a lot of money to set up labs and to attract first class researchers from around the globe. One of the concerns during the Bush Administration (2001–2008) was that the limitations on federal funding for hESC would lead to a brain drain of scientists and disadvantage the United States in this booming field.[66] There is no evidence that this brain drain took place. While it is true that stem cell researchers are mobile individuals, it is more likely that many engaged in collaborative research with their counterparts in other countries. This is easy to discern when looking at the affiliations of a research team. Research scientists spend months or even years as postdoctoral researchers in the labs of well-known scientists to learn techniques.

Although there was less political enthusiasm in the United States for hESC research funding, the field continued to expand. Thomson and Gearhart had both concluded their breakthrough papers by hypothesizing that research using hESC could lead to lifelong treatments for many diseases, such as Parkinson's disease and juvenile-onset diabetes. Both used the words "transplantation medicine," which have today become a key phrase in the field of regenerative medicine. The hope was that these therapies were just around the corner. What both scientists stated clearly in their papers was that, while the research done with mouse embryos had been able to differentiate the cells (that is, turn the stem cells into different types of tissue), this had not yet been done with the hESC lines. This would occur in time, but in 1998 the research was still in its early stages. This did nothing to dampen the explosive interest in the field over the next decades.

Research interest in hESC and adult stem cells took off in the twenty-first century. One indication of this growth was the formation of the International Society for Stem Cell Research (ISSCR) in 2001. It began sponsoring the journal *Cell Stem Cell*, considered by some to be one of the top journals in the field.[67] The organization now has over 4000 members from 60 countries and holds annual international and regional meetings. Moreover, it is a vehicle for specialized training for members and nonmembers. If you want to get a job in a stem cell lab but do not have time to get an MD and/or a PhD, then a certification in single cell analysis or mouse development might be an alternative.

New journals in this field have also proliferated. In 2006, the *Journal of Stem Cells and Regenerative Medicine* published its first volume. In 2010, the *Journal of Stem Cell Research and Therapy* hit the ground. In 2014, the *Journal of Stem Cell Research and Transplantation* started publication. There are many more open-source, online journals that have emerged, and many of their titles indicate a focus on transplantation or regenerative medicine.

The European Union launched its stem cell hub (EuroStemCell) in 2004 to help its citizens make sense of stem cell research. The hub is membership-based but also receives funding from the European Union. It publishes a newsletter and special studies. A primary goal of the center is to educate citizens about potential stem cell treatments.

Another quickly growing sector related to stem cell research is institutes for regenerative medicine. While most large research universities have many different specialized science labs, the multidisciplinary institute is becoming more common. We cite just one example that tends to serve as a prototype: the McGowan Institute for Regenerative Medicine at the University of Pittsburgh. Originally founded in 1992 as the McGowan Center for Artificial Organ Development, it was reorganized and renamed in 2001 when its mission was expanded to include tissue engineering and adult-derived stem cells. The backgrounds of the more than 200 researchers who work for or are affiliated with the institute underline the emphasis on translational medicine. The staff includes engineers, physical therapists, medical doctors, and cell biologists. It seems evident that moving stem cells out of the lab can only be facilitated by multidisciplinary teams working on a project.

Not to be outdone by academia, we also find increased interest in the private sector. Old and new companies are moving into the stem cell arena. We have already mentioned Geron, which was founded in 1990 and continues to be a leader in this research area. New start-up companies emerged in states that already had a reputation for supporting stem cell research. California and Massachusetts are two states that promoted and funded hESC research. California became an especially desirable location after the state passed a $10 billion initiative in 2004 to create the California Institute for Regenerative Medicine (CIRM). Some companies are started by the scientists themselves after they develop the products or procedures, receive patents, and find financial backers; other new companies supply tools and researchers to move discoveries from clinic to market. A number of companies specialize in assays, or the testing of the cell-based therapies or drugs. This sort of business operation resembles contracting out blood analysis.

The increase in the number of scientists and institutions doing work in the field of stem cell research has had results both positive and negative. The positive result is that, if more people are working on similar research (or with the same cell tissues), new and more effective findings are likely to emerge. Similarly, if there are more products/therapies in clinical trials, there is a better chance that products will reach the market.

On the other hand, growth in the field may also have contributed to incidents of fraud and deception. Two types of fraud have emerged, and it is arguable which is most egregious. The first is researcher fraud, with scientists faking their data and results. The second is product fraud, which contributes to stem cell tourism.

Induced pluripotent stem cells (iPSC): Avoiding ethical concerns?

While hESC cells are still considered the gold standard in stem cell research, there were a number of problems with using embryonic cells. Ethically, the act of destroying an embryo was considered tantamount to destroying a life. This issue will be discussed in greater detail in Chapter 5. Other practical problems included getting high-quality embryos, improving on the media used for growing cells, and getting the cells to differentiate into the desired tissues.[67] Another practical issue, when suitable therapies were developed, was rejection by the host (that is, the patient) due to tissue incompatibility. If researchers could find a way to use the patient's own cells to create therapies, this would solve any rejection issues.

The discovery in 2006 of iPSC by a research team in Japan was a major breakthrough that earned Shinya Yamanaka a Nobel Prize in 2012.[68] Yamanaka and his team, first using mouse cells and later using human adult skin cells, used four transcription factors (proteins) and injected them into an adult cell, causing the cell to return to an embryonic-like state (See Figure 2.5). These cells could then be differentiated into different germ layers.

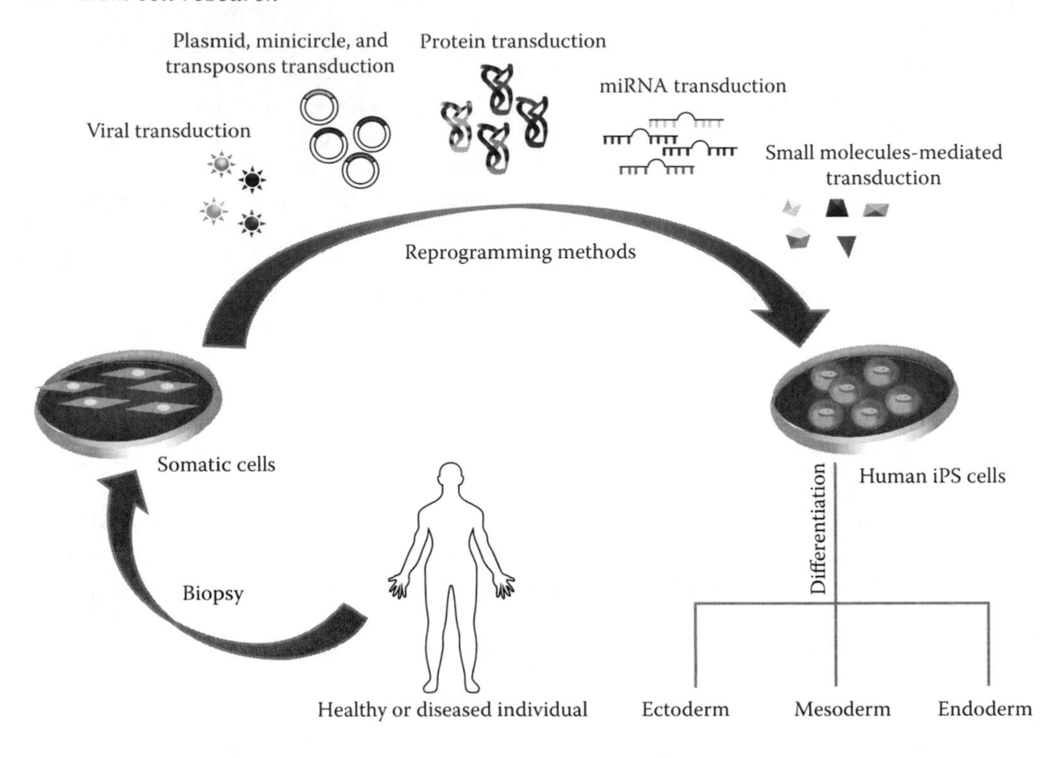

Figure 2.5 Creation of Induced Pluripotent Stem Cells (Reproduced from Biochemj.org, http://www.biochemj.org/content/465/2/185.)

This finding constituted an important turning point in the field as more researchers focused first on replicating Yamanaka's research,[69] and then on trying to find different transcription factors that would speed up the differentiation process.[70] One area of research that is attracting a lot of attention is trying to eliminate the middle stage of turning the adult cells into pluripotent cells and then into different specialized cells. By avoiding the middle step, cells can be generated more efficiently, saving time and money. Research teams at Stanford and Columbia universities[71] have had some success in turning skin cells into neuron cells (brain), which could be used to derive therapies for Alzheimers's, Parkinson's, and other brain-related diseases. This is a very fertile area of research, especially with the new crop of stem cell researchers who have genetic and pharmacological training.

While iPSCs look like a promising method for deriving specialized cells for disease modeling, the ongoing concern is that iPSCs could develop different characteristics (that is, gene expression differences) after a few passages in a Petri dish when compared to hESCs.[72] This difference aside, iPSCs are being used by researchers to model diseases and speed up drug development. What researchers are looking for are faster and cheaper ways to create these iPSCs.

Negative consequences: Fraud and tourism

The negative consequences of the growth and interest in this new field are the incidents of researcher fraud that cast a negative light on the validity of scientific findings. A second unfortunate and possibly more dangerous consequence is the increase in unproven treatments sold to patients as stem cell therapies and usually administered outside the United States. The second is referred to as stem cell tourism.

While many of us have great respect for the work done by scientists, it is also clear that in a growth field like stem cell research, some scientists may be cutting corners to get their research published. Recall that we said the basis of the scientific method is replication. If other scientists cannot recreate the same results, then the findings are suspect.

Researcher fraud

Two cases of stem cell fraud that drew international attention occurred in South Korea and Japan. In the case of Hwang Woo-suk, a well-known South Korean scientist, we suspect that he was interested in retaining his star status in Korea and abroad. In 2004[73] and 2005, his lab published two papers claiming to have created a line of hESCs by using somatic cell nuclear transfer (remember Dolly). They transferred the nucleus of an adult cell into an unfertilized egg. Given that Hwang was an expert in cloning, the findings did not seem suspicious. However, in less than 6 months the truth was uncovered: Hwang and his team had faked the data and had also obtained human egg donations under false pretenses. The scientific community began some serious soul searching. What had gone wrong with the peer review process? Was it possible that in a large lab such as Hwang's there was no oversight? Did junior researchers, hoping to please their director, alter the findings? Hwang was fined and removed from his post at Seoul National University, but South Korea's reputation as a stem cell leader had vanished. The incident cast a pall over the stem cell research field. Some attributed the scandal to an increased pressure on scientists to publish (you might recall the increase in the number of journals in this field), coupled with the inability of the peer review process to detect fraud. Another possible contributing factor is that many scientific discoveries are patented. Biotechnology companies play an important role in funding but also marketing medical discoveries (think of Geron). Korea has a significant biotechnology sector, and Hwang's research and international reputation helped promote investments in biotech companies.[74,75]

In 2014, another equally surprising scientific discovery turned out to be too good to be true. Japanese researchers at the RIKEN Center for Developmental Biology published two papers revealing they had discovered a way to produce iPSCs by simply subjecting adult cells to stressors such as low oxygen levels and acidic environments.[76] The procedure was called stimulus-triggered acquisition of pluripotency (STAP), and for a few months this procedure was thought to be a major breakthrough for a young Japanese researcher named Haruko Obokata, who became famous overnight. Both papers were coauthored by U.S. researchers from Boston's Brigham and Women's Hospital and Harvard University, so at first, it did not seem that anything was amiss.

If the research was accurate, the procedure would make it possible to create patient-specific stem cell therapies relatively easily. The procedure did not require nuclear transfer, like the Hwang research, or the introduction of transcription factors, as with the Yamanaka method. Once the adult cells were stressed, they turned into embryonic-like stem cells. These cells could then be differentiated into various cell and tissue types. However, once the articles were published, researchers around the globe were unable to replicate the research. The articles were retracted due to multiple errors, but Dr. Obokata was given 8 months to try to replicate her results. She was unable to do so. This incident was accompanied by a series of personal tragedies. The codirector of Japan's prestigious RIKEN Center and Obokata's mentor committed suicide, the director resigned, and Obokata was charged with research misconduct.[77] Once again, the research community began to discuss the factors that contribute to scientific fraud. The factors that emerged are similar to those discussed after the Hwang case—increased competition in the field, pressure to publish, and the desire to make the next big discovery.

While these instances of research misconduct became international events, there may be others that exist without garnering the same level of attention. One red flag is the retraction of an article published months or years ago. Is this fraud or just sloppy research? It may be hard to determine. But surely such an occurrence calls into question the integrity of the research group.

Stem cell tourism

Medical tourism occurs when a person leaves his or her home country for treatment abroad because (a) the treatment/drug is not approved for use in the home country, or (b) the treatment costs less in another country. When people go abroad for face-lifts, hip replacements, or root canals, we regard this as a personal choice. However, when patients go abroad for untested treatments, there is concern for the safety of the patients, the reputation of the medical profession, and the integrity of the country/company providing the treatment.

In the United States, all drugs and drug clinical trials must be approved by the Food and Drug Administration (FDA). Stem cell therapies are treated as drugs. The FDA[78] and the ISSCR[79] provide information for patients who are thinking about going abroad for stem cell treatment. The NIH also lists all of the clinical trials being conducted in the United States and abroad.[80] The NIH makes it clear that it is not responsible for the safety or effectiveness of out-of-country treatments.

All of these warnings notwithstanding, patients, especially those with life-threatening conditions, are enticed by Internet ads complete with testimonials from satisfied patients claiming that the treatment works. A recent article estimated that between 60,000 and 750,000 people go abroad annually seeking treatments.[81] More than half of all stem cell clinics are based in developing countries. It is difficult to monitor, let alone regulate, clinics engaged in trial therapies. It is only when patients see no benefits that they are likely to complain about the treatment. Patients in Canada recently complained to authorities about a company that recruited multiple sclerosis patients in Canada and sent them to India for treatment. Patients paid to participate (legitimate clinical trials require no payment) and saw no lasting results. Patients are now complaining that the Canadian government should do something about the company.[82] In the United States, a company that offered adult stem cell therapies in Texas was forced to move its operation to Mexico after receiving a warning from the FDA that its therapies were not approved. Given that one of the company's patients had been Texas Governor Rick Perry, perhaps the company felt it could fly under the radar.[83]

In 2016, China (a favorite destination, along with India, for questionable stem cell therapies) issued laws to regulate stem cell research and crack down on clinics offering untested cures.[84] China-watchers are skeptical that the new laws will have any lasting results.[85]

There is no dearth of clinics and companies that seek to take advantage (financially and scientifically) of people looking for cures or treatments for debilitating conditions. It is difficult for countries to regulate and penalize such companies. Even in the United States, where there is a strong regulatory tradition, these predatory companies are sometimes able to avoid detection. As we will discuss in the next section, there are only a few small, approved clinical trials using hESCs and one using iPSCs as of this writing.

Clinical trials: Safety first

In the United States, all drugs, including stem cell therapies, must receive approval from the Food and Drug Administration. A drug is "any product that is intended for use in the diagnosis, cure mitigation, treatment, or prevention of disease and that is intended to affect the structure or any function of the body."[86]

Clinical trials follow a series of four distinct phases. The first phase is intended to determine if the drug/therapy is safe. In this phase, a small group of people will be used as a pilot. Depending on the study, the group might consist of healthy subjects or participants who have the condition/disease that the therapy is designed to improve. Phase 2 (usually phases 1 and 2 are combined) entails giving the therapy to a larger group, and the efficacy of the treatment is also monitored. The participants might receive a higher dosage of the drug/treatment. Phase 3 is designed to expand the number of participants, monitor side effects, and confirm safety and effectiveness. Phase 4 expands the trial to an even larger population and is often initiated after the drug or treatment is on the market. The safety, effectiveness, and side effects of long-term usage are studied.[87]

This is a long and costly process and may explain, in part, why there are so few clinical trials using hESC therapies and why all of the trials are in phases 1 and 2. Secondly, while hESC (and iPSC) therapies have great potential, getting these cells to differentiate in a reliable manner continues to be an issue. Trial therapies are tested on mice first, but the move from a mouse model to a human model does not guarantee safety. In fact, even moving a therapy from men to women may not always be safe.

Spinal cord injury (SCI)

According to the Christopher and Dana Reeve Foundation, six million people suffer from an SCI, and the associated estimated yearly expenses can range from $238,000 to $750,000.[88] This is one of the reasons why researchers were eager to find a workable treatment. It should also be mentioned that Christopher Reeve (paralyzed in an equestrian accident in 1995) became a vocal advocate of stem cell research.

In 2009, Geron and the University of California, Irvine, received FDA approval to begin a phase 1 trial using hESCc for spinal cord injury. The study was delayed a year due to some concerns over the purity of the cells to be injected.[89]

Geron turned hESCs into neural support cells (oligodendrocyte progenitor cells) that were injected into patients with the hope that this treatment would restore some patient sensation. There was no expectation that the patients would be able to walk again as the mice had in initial studies.[90] Patients who enrolled in the trial needed to have a complete injury with virtually no chance of spontaneous recovery. This was to ensure that the therapy was making a difference.

The trial was approved for ten participants, but only five were enrolled before Geron abandoned the research due to financial issues in 2011. Of the five people treated, none had serious side effects—so the safety condition was met—but none demonstrated any major improvement in mobility or sensation.

In 2013, Asterias Biotherapeutics assumed the right to continue the trial with an altered protocol. The FDA approved a phase 1 and 2 clinical trial in 2014 to enroll 13 participants who were 14–30 days post injury. The first three patients enrolled were given low doses of the test cells (about 2 million cells); subsequent patients would be given higher doses (10–20 million cells).[91] The expectation was that patients would see improvements in mobility. The first three patients enrolled showed no adverse effects, but there is no report as of this writing that their mobility/sensation improved. But the goal of a phase 1 and 2 clinical trial is safety. The only adverse effects reported were due to the immunosuppressant drugs that patients were required to take.

Asterias is still enrolling participants, and it is possible that with a larger sample and larger doses of hESCs, positive results will emerge. Asterias received a $14 million grant from the California Institute for Regenerative Medicine to continue the research. There is definitely a lot of interest in the outcome of this trial, but the costs of expanding the study are high. The goal is not that patients will get up and walk again, but rather that their mobility will be improved so that they do not require continuous care. With many of the conditions for which scientists are trying to find treatments, it is the cost of care for the patient that is driving the research. Most individuals who develop Alzheimer's require around-the-clock care. This level of care is very expensive financially if given by a commercial provider and very expensive psychologically if given by a family member.

Age-related macular degeneration and Stargardt disease

Age-related macular degeneration (AMD) is one of the leading causes of vision loss in people over 65, according to the National Eye Institute. The Institute also estimates that the number of people with the disease will double by 2050 to 5.5 million people. Stargardt macular degeneration, on the other hand, is a genetic eye disorder and a leading cause of juvenile blindness.[92]

Two clinical trials, one in the United States and one in England, are using hESCs to treat one or both of the diseases. A third trial in Japan is using iPSCs to treat AMD. The Japanese study is the first to use iPSCs. In the United States, Advanced Cell Technology (now called Ocata Therapeutics) received FDA approval in 2011 for two clinical trials, one for AMD that enrolled nine people, and one for Stargardt disease that also enrolled nine people. After 2 years, 10 out of the 18 participants reported improvements in vision. The therapy seems to have halted the progression of the disease. Ocata plans to expand the clinical trial and to increase the dosage of stem cells administered. The main goal of the study was to show that the process is safe.[93]

In 2015, the London Project to Cure Blindness, with financial help from the New York Stem Cell Foundation, CIRM, and Pfizer Inc., began clinical trials for people with wet-eye AMD. In this instance, cells are grown in a lab to form a patch that is placed behind the retina during surgery. This trial is also in phases 1 and 2 to determine safety and efficacy. As of this writing, only one patient has undergone treatment, but up to 10 patients will be recruited. Drs. Pete Coffey and Lyndon da Cruz, who are the cofounders of the London Project, are optimistic that the treatment can be modified and used on patients suffering from Stargardt's disease and also dry AMD.[94]

The RIKEN Center for Developmental Biology in Japan enrolled its first of two patients in 2014. The cells were derived from the patient's own cells (an autologous stem cell transplant). The trial was suspended in 2015 when genetic mutations were discovered; it was recently announced that this trial is scheduled to resume in 2017; the road from discovery to delivery often takes longer than expected (or hoped for).[95] The new trial will be a collaboration between RIKEN and Kyoto University, where Shinya Yamanaka (the scientist who discovered iPSCs) is head of the Center for iPS Cell Research and Application. This trial will involve allogenic cells (derived from a different patient).

Diabetes

According to the American Diabetes Association, 29.1 million Americans had diabetes in 2012, and the number is projected to go up. Using hESC to find a cure became the life goal of one Harvard researcher, Douglas Melton, who is the father of two children with juvenile diabetes. He succeeded recently in coming up with a way to create cells that could be transplanted into patients.[96] At the same time, ViaCyte Inc., a San Diego-based biotechnology company working in the same area, received FDA approval in 2014 to begin a clinical trial using a similar model for implanting cells into patients. The ViaCyte trial is listed by the NIH as recruiting participants for test sites in both the United States and Canada.[97] The study has just commenced.

Drug discovery and personalized medicine

While there is some activity in human clinical trials in the United States and England, it is not the groundswell that some might have expected after 18 years of research with hESCs. As research moves from the lab to the clinic and beyond, it is expected that more therapies will emerge for other debilitating and fatal diseases. Some of these trials will, in time, make it to market. It takes time for new therapies to become clinically useful and routine practice. It took 30 years for bone marrow transplantation to become routine. The two areas that seem most promising for stem cell research and application are drug discovery and personalized medicine.

Drug development is a costly activity, as pharmaceutical companies are quick to point out when the price of a medication is questioned. A recent study by the Tufts Center for the Study of Drug Development put the cost of developing a new drug at $2.6 billion.[98] This estimate may be a bit high, but new drugs are typically tested first in animals, followed by clinical trials with humans. Add to this cost the liability that companies assume when a drug does make it to market, and we get a better picture of why drugs cost so much. How can stem cells help speed up this process and lower costs?

Growing hESC and iPSC in a dish, getting them to differentiate into specific tissue types, and then testing new drugs on the different tissue types is an efficient way to determine safety. The discovery of iPSC cells in 2006 makes testing diseases in a dish less controversial (no embryos are destroyed) and more efficient. Also, since adult cells are used to produce these cells, it is possible to combine drug testing with personalized medicine.

Heart stem cells in a dish are becoming more common as a way to test whether a drug is likely to have an adverse impact on a patient. Had this type of testing been available in the 1990s when Merck tested its popular but dangerous drug Vioxx, it might have discovered that the painkiller was also a heart stopper. The drug was withdrawn in 2004 after several patients suffered heart attacks and strokes.[99] Cardiotoxicity (the side effect of a drug on the heart) is an important required element of drug testing following the Vioxx incident. The other benefit of disease testing in a dish is that cells can be grown that have the disease the drug is targeting. There seems to be a lot of activity in this area as pharmaceutical companies look for cheaper and more effective ways to test new drugs. In addition, the approval process for testing drugs in a dish is far shorter than the process for getting a human clinical trial approved by the FDA.

Patient-specific stem cell treatment is another area that looks very promising since the discovery of iPSCs. The idea of personalized treatment or individualized medicine is not new. The journal *Personalized Medicine* was founded in 2004 and taps into the field of personalized genomic medicine.[100] By creating patient-specific iPSCs, treatments and drugs can be tested in a dish before being used on the patient. While personalized medicine is still an experimental and costly method of improving patient health, it is, nonetheless, an important area of research and one in which stem cell research is playing an active role.

Summary

Human embryonic stem cell research is a new field of scientific inquiry. In this chapter we have seen the development of the field and the debt it owes to a number of other medical discoveries. It took decades for doctors to develop predictable protocols for bone marrow transplants, which are now well-established procedures. This is the future that many hope will arrive for hESC and iPSC research. We have a few more decades to go before the discovery becomes a standard procedure and the era of personalized or precision medicine is ready for its prime.

Additional readings

Original sources and other scholarly readings

Scholarly research requires reading original sources in addition to secondary sources. Some of the articles in this section are very scientific. Test your skill at thinking like a cell biologist.

1. J. Gearhart, New potential for human embryonic stem cells, *Science* 282 (5391), November 1998: 1061–1062; http://www.ncbi.nlm.nih.gov/pubmed/9841453.
2. J. A. Thomson et al., Embryonic stem cell lines derived from human blastocysts, *Science* 282 (5319), November 1998: 1145–1147; http://www.ncbi.nlm.nih.gov/pubmed/9804556.
3. M. J. Shamblott et al., Derivation of pluripotent stem cells from cultured human primordial germ cells, *Proceedings of the National Academy of Sciences* 95, November 1998: 13726–13731; http://www.ncbi.nlm.nih.gov/pubmed/9811868.

4. NIH, *The Human Embryonic Stem Cell and the Human Embryonic Germ Cell*, http://stemcells.nih.gov/info/scireport/pages/chapter3.aspx [A timeline of important discoveries and contributions in the development of stem cells].
5. NIH, *Learn about Clinical Trials*, https://clinicaltrials.gov/ct2/about-studies/learn.
6. NIH, Stem cell basics, http://stemcells.nih.gov/info/basics/pages/basics3.aspx.
7. P. S. Knoepfler, Key anticipated issues for clinical use of human induced pluripotent stem cells, in *Regenerative Medicine,* 7 (5): September 2012: 713–720 http://www.ncbi.nlm.nih.gov/pubmed/2283062.
8. S. Minger, Blue Skies: The future of regenerative medicine, February 1, 2015, at https://www.youtube.com/watch?v=w5Yohe2jZd4 [Stephen Minger, Chief Scientist, Life Sciences, GE Healthcare].
9. S. Solomon, *Realizing the Promise of Stem Cell Research*, at http://nyscf.org/susansolomontedtalk; https://www.ted.com/talks/susan_solomon_the_promise_of_research_with_stem_cells?language=en [Susan Solomon is the co-founder and CEO of the New York Stem Cell Foundation].
10. Howard Hughes Medical Institute, *Potent Biology: Stem Cells, Cloning and Regeneration, 2006 Holiday Lecture*, Douglas Melton, Lecture 1: Understanding Embryonic Stem Cells; https://www.hhmi.org/biointeractive/understanding-embryonic-stem-cells.

Secondary analysis and news articles

1. A. Wolfe, Susan L. Solomon's stem-cell research quest, *Wall Street Journal*, February 5, 2016: http://www.wsj.com/articles/susan-l-solomons-stem-cell-research-quest-1454699397.
2. E. Yong, Testing drugs on Mini-Yous, Grown in a Dish, *The Atlantic*, June 22, 2016; http://www.theatlantic.com/science/archive/2016/06/testing-drugs-on-mini-yous-grown-in-a-dish/488039/.
3. J. A. Johnson and E. D. Williams, Stem cell research, *Congressional Research Service Report for Congress,* Updated August 10, 2005, at http://fpc.state.gov/documents/organization/51131.pdf.
4. 4. C. Dreifus, At Harvard's stem cell center, the barriers run deep and wide, *New York Times*, January 24, 2006; http://www.nytimes.com/2006/01/24/science/at-harvards-stem-cell-center-the-barriers-run-deep-and-wide.html?_r=0.

Critical thinking activities

1. Before new drugs or therapies can be given to the general patient population, they must undergo a clinical trial. The National Institutes of Health lists all of the active clinical trials. Go to the NIH website: https://clinicaltrials.gov/ct2/about-studies/learn. Take a few minutes to read the different sections that tell you about who conducts the studies, how long a trial can take, who can participate, and so on. You can also check out other sites that discuss clinical trials at https://clinicaltrials.gov/ct2/about-studies/other-sites. Now go to: https://clinicaltrials.gov. This is the NIH home page for finding all the active clinical trials in both the United States and other countries. You can do a basic search from this site—if you put in, for example, "spinal cord injury" or "spinal cord injury" and "California," you will get all the trials being conducted on this condition in California. You can do an advanced search at https://clinicaltrials.gov/ct2/search/advanced. This allows you to put in a search term, such as "stem cells" or "human embryonic stem cells," and a condition, such as "spinal cord injury" or "Alzheimer's." For this assignment, you are to:
   a. Search for three clinical studies using the search terms "stem cells," "human embryonic stem cells," or "induced pluripotent stem cells" (you might want to narrow the stem cells to heart cells or brain cells) and some disease. Pick a disease or condition you are interested in learning more about. HINT: After reading this chapter, you

know there are few trials using hESC, so that might not be a great search term. But critical thinking is about testing your ideas and hypotheses, so you might find a trial in another country. Not all of the trials that are listed will be actively recruiting, but you can still find out what the trial was trying to do.

b. Read about the three studies you have selected, focusing on recruitment timeframe, eligibility for participation, trial location, and any other conditions listed.
c. Write a five-page essay in which you compare and contrast the requirements for participation. What do the trials hope to achieve? Assume that you have the condition/disease being investigated; would you be willing to participate? Why or why not? Note that you may select at least one study that is being conducted outside the United States. Typically, about half of all trials are located outside the United States.
d. Write a five-page paper in which you discuss why it is safer and more cost-effective to focus on this line of research rather than trying to find therapies to cure individual patients. Note that you may want to do some research on how much it costs and how long it takes to bring a drug to market. And even when it gets to your pharmacy, it may not be safe for you. Before clinical trials are done on humans, most drugs and therapies are tested on animals.

2. Human embryonic stem cells, since they were first isolated, have received a lot of attention. Scientists, politicians, and patient advocates point to their potential for curing many of the most debilitating and costly diseases. But one area of research that is not getting as much press coverage but may provide more immediate results is drug screening. By using in-vitro human tissues grown from stem cells, scientists have an opportunity to assess the toxicity of drugs and other substances on human cells rather than on animals. Listen to Susan Solomon, the cofounder and CEO of the New York Stem Cell Foundation talk about research with stem cells at either of the following sites: http://nyscf.org/susansolomontedtalk or https://www.ted.com/talks/susan_solomon_the_promise_of_research_with_stem_cells?language=en. Also listen to Stephen Minger, chief scientist for cellular sciences at GE Healthcare Life Sciences, talk about the future of regenerative medicine. His lecture is a bit longer, but you will see heart cells beating in a dish: https://www.youtube.com/watch?v=w5Yohe2jZd4.
3. Is it possible to understand the science in an hour with the help of one of the top stem cell researchers in the United States? Listen to Dr. Douglas Melton, professor of stem cell and regenerative biology at Harvard University and codirector of the Harvard Stem Cell Institute, give a video lecture on *Understanding Embryonic Stem Cells.* After watching the lecture, pick two of the segments in which he discusses some of the issues that continue to puzzle scientists (the lecture is divided into about 40 substantive subsections). After doing some additional research, write a three-page paper in which you discuss why scientists are still trying to figure out some of the key issues that he brings up. HINT: Consider differentiation of hESCs. The lecture can be viewed from the following websites: https://www.hhmi.org/biointeractive/understanding-embryonic-stem-cells or https://www.youtube.com/watch?v=nYNBNZJ8Xck.

Notes

1. J. Slack, *Stem Cells: A Very Short Introduction,* New York: Oxford University Press, 2012, 6. According to Slack, the term is more common among journalists and politicians. The National Institutes of Health glossary refers users to the term "somatic stem cells." In this book, we will use the term adult stem cell, except when it is necessary to indicate in what tissue the cell is located. For example, we may refer to blood stem cells or skin stem cells.
2. A totipotent cell is produced when a sperm and egg cell unite. The first few divisions produce more totipotent stem cells. However, embryonic cells within the first couple of cell divisions after fertilization are the only ones that are totipotent. After 4 or 5 days of embryonic cell division, the cells begin to specialize. On the fourth or fifth day of development, the embryo forms into two

layers: an outer layer that will become the placenta and an inner layer mass that will form all the tissues of the developing human body. The cells taken from a human embryo to form a stem cell line are taken from this inner layer of cells. The National Institutes of Health provides a diagram depicting this process: www.stemcells.nih.gov/info/basics/pages/basics3.aspx.

3. J. Gearhart, New potential for human embryonic stem cells, *Science* 282, November 1998: 1061–1062; J. A. Thomson et al., Embryonic stem cell lines derived from human blastocysts, *Science* 282, November 1998: 1145–1147; M. J. Shamblott et al., Derivation of pluripotent stem cells from cultured human primordial germ cells, *Proceedings of the National Academy of Sciences* 95, November 1998: 13726–13731.
4. See the National Institutes of Health website for a timeline of important discoveries and contributions in the development of stem cells. National Institutes of Health, Appendix C: Human Embryonic Stem Cells and Human Embryonic Germ Cells; https://stemcells.nih.gov/info/2001report/appendixC.htm.
5. The NIH National Library of Medicine, MedlinePlus, defines graft-versus-host disease (GVHD) as "a complication that can occur after a stem cell or bone marrow transplant." The donated blood or cells view the recipient's body as foreign and begin to attack the body. Today, the Department of Health and Human Services (as well as private hospitals) maintains a database on bone marrow stem cell transplants. These organ procurement and transplantation networks help locate suitable donors to minimize the graft-versus-host problem. See Glossary for definition.
6. A. B. Parson, *The Proteus Effect: Stem Cells and Their Promise for Medicine,* Washington DC: Joseph Henry Press, 2004; A. Park, *The Stem Cell Hope: How Stem Cell Medicine Can Change Our Lives*, New York: Hudson Street Books, 2011; D. Baron, The global race for stem cell therapies, *BBC News*, April 26, 2005.
7. P. S. Knoepfler, Key anticipated regulatory issues for clinical use of human induced pluripotent stem cells, in *Regenerative Medicine* 7, September 2012: 713–720. The cost of stem cell therapies is rarely mentioned, as few data are readily available. However, Knoepfler does mention that costs might be higher than anyone expects. See also, K. Narioka and P. Dvorak, Japan makes advances on stem cell therapies, *Wall Street Journal* 10, June 27, 2013: 116. The *Wall Street Journal* article quotes a leading stem cell researcher, Masayuki Yamato, who said initial treatments for age related macular degeneration could be as high as $512,000 per person. It is not clear who would pay for these therapies. The stem cell therapies that are currently being tested are experimental, with the hospital or research organization assuming most of the cost. But once these therapies become routine, it is not clear how the costs will be determined or distributed. Will medical insurance cover the cost?
8. Stem cell tourism is a big issue now that more therapies (proven or not) become available in countries where medical controls are less restrictive than in the United States. Both the National Institutes of Health and the International Society for Stem Cell Research (ISSCR) have advice to patients on their websites suggesting questions for patients to ask before engaging in therapies. International Society for Stem Cell Research, *ISSCR Patient Handbook on Stem Cell Therapies*, 2008. The handbook can be downloaded from http://www.isscr.org/home/publications/patient-handbook. See also, A. D. Levine and L. E. Wolf, The roles and responsibilities of physicians in patients' decisions about unproven stem cell therapies, *Journal of Law, Medicine & Ethics* 40, Spring 2012: 122–134. National Public Radio (NPR) also did a program in 2010 discussing clinics that offer different kinds of stem cell therapies: R. Knox, Offshore stem cell clinics sell hope, not science, National Public Radio, 2010.
9. B. Alberts et al., *Essential Cell Biology*, New York: Garland Science, 2013.
10. National Institutes of Health, National Human Genome Research Institute, *Fact Sheets on Science, Research, Ethics and the Institute*; https://www.genome.gov/10000202/. NIH provides a number of examples and explanations regarding what the Human Genome Project is and what it can do. It is written for the nonscientist, with references to more scholarly articles by leading researchers for those who might be interested. Most texts on biology and genetics also have step-by-step explanations as to how scientists use the information.
11. H. M. Schmeck, Jr., DNA and crime: Identification from a single hair, *New York Times*, April 12, 1988; also, E. Eckholm, No longer ignored, evidence solves rape cases years later, *New York Times*, August 2, 2014.
12. Diagrams of the four main tissue types can be seen at: United States Department of Health and Human Services, National Institutes of Health. Students who are interested in learning more about

the four tissue types should see: K. T. Patton and G. A. Thibodeau, *The Human Body in Health and Disease*, Atlanta, GA: Elsevier, 2013; also, R. Phillips et al., *Physical Biology of the Cell,* New York: Garland Science Publishing, 2013.

13. NIH, *Stem Cell Basics*, April 8, 2015; http://stemcells.nih.gov/staticresources/info/basics/SCprimer2009.pdf.
14. S. Boseley, First UK patient receives stem cell treatment to cure loss of vision, *The Guardian,* September 29, 2015. The research is under the guidance of Drs. Pete Coffey and Lyndon Da Cruz. Both are affiliated with the London Project to Cure Blindness. They have been working for more than 10 years on this project and hope to see clinical results soon.
15. L. K. Altman, Test tube skin helps save 2 burn victims, *New York Times*, August 16, 1984; also, A. Kneller, Stem cell treatment for burn patients earns alpert prize. *Harvard Medical School News,* September 3, 2010; http://hms.harvard.edu/news/stem-cell-treatment-burn-patients-earns-alpert-prize-9-3-10; also, G. J. Todaro and H. Green, Qualitative studies of the growth of mouse embryo cells in culture and their development into established lines, *Journal of Cell Biology* 17, May 1963: 299–313. In 2010, Howard Green published a book (more of a memoire) of how he and his colleagues spent 14 weeks working around the clock to culture enough skin cells to save the lives of two young burn victims. H. Green, *Therapy with Cultured Cells*, Singapore: Pan Stanford Publishing, 2010.
16. NIH, *All About the Human Genome Project*; https://www.genome.gov/10001772/all-about-the--human-genome-project-hgp/.
17. K. Takahashi and S. Yamanaka, Induction of pluripotent stem cells from mouse embryonic and adult fibroblast cultures by defined factors, *Cell* 126, August 2006: 663–676.
18. J. A. Thomson et al., Embryonic stem cell lines derived from human blastocysts, *Science* 282, November 1998: 1145–1147; also, M. J. Shamblott et al., Derivation of pluripotent stem cells from cultured human primordial germ cells, *Proceedings of the National Academy of Sciences* 95, November 1998: 13726–13731.
19. One team trying to grow human embryonic stem cell lines was headed by Ariff Bongso from Singapore. A. Bongso et al., Fertilization and early embryology: Isolation and culture of inner cell mass cells from human blastocysts, *Human Reproduction* 9, November 1994: 2110–2117. Among the problems that stumped the research was (a) getting good quality embryos from which they could extract a cell; and (b) finding a suitable medium on which to grow the cells, and, (c) making sure that researchers separated the cell before they started to differentiate/specialize. In her comprehensive book, C. Fox, *Cell of Cells: The Global Race to Capture and Control the Stem Cell*, New York: Norton, 2007, discusses the numerous research groups around the world that were working on trying to grow human embryonic stem cells. In short, keeping the embryos pluripotent was not an easy job. Appendix 2 of Michael Bellomo's book includes the rather elaborate protocol from Wake Forest University on the steps needed to ensure that embryonic stem cells were grown properly. M. Bellomo, *The Stem Cell Divide: The Facts, the Fiction, and the Fear Driving the Greatest Scientific, Political, and Religious Debate of Our Time*, New York: AMACOM, 2006.
20. A. M. Stewart and G. W. Kneale, A-Bomb Survivors: Factors that may lead to a reassessment of radiation hazard, *International Journal of Epidemiology* 20, January 2000: 708–714; also, P. Voosen, Hiroshima and Nagasaki cast long shadow over radiation, *New York Times*, April 11, 2011.
21. B. Deschler and M. Lubbert, Acute myeloid leukemia: Epidemiology and etiology, *Cancer* 107, October 3, 2006: 2099–2107.
22. E. A. McCulloch, and J. E. Till, The radiation sensitivity of normal mouse bone marrow cells, determined by quantitative marrow transplantation into irradiated mice, *Radiation Research* 13, July 1960: 115–125; also, J. E. Till and E. A. McCulloch, A direct measurement of the radiation sensitivity of normal mouse bone marrow cells, *Radiation Research* 14, February 1961: 213–222.
23. A. J. Becker, E. A. McCulloch, and J. E. Till, Cytological demonstration of the clonal nature of spleen colonies derived from transplanted mouse marrow cells, *Nature* 197, February 1963: 452–454.
24. U.S. Department of Health and Human Services, Health Resources Service Administration (HRSA) maintains a database that lists all of the more than 200 centers in the United States that perform transplants. In 2013, more than 5000 transplants were performed, HRSA, *U.S. Transplant Data by Center*; http://bloodcell.transplant.hrsa.gov/research/transplant_data/us_tx_data/data_by_center/center.aspx.

25. L. Lasky, Cord blood and our tomorrow, *American Association of Blood Banks: News*, March–April 2001; also, R. M. Kline, Whose blood is it anyway? *Scientific American* 284, 2001: 42–49.
26. R. Waller-Wise, Umbilical cord blood information for childbirth educators, *The Journal of Perinatal Education* 20, 2011: 54–60; also, D. Drew. Umbilical cord blood banking: A rich source of stem cells for transplant, *Advance for Nurse Practitioners* 4, 2005: S2–S7; American Academy of Pediatrics, Policy Statement: Cord blood banking for potential future transplantation, *Pediatrics* 119, January 2007: 165–170.
27. G. J. Todaro and H. Green, Quantitative Studies of the growth of mouse embryo cells in culture and their development into established cell lines, *Journal of Cell Biology* 17, May 1963: 299–313; J. G. Rheinwald and H. Green, Serial cultivation of strains of human epidermal keratinocytes: The formation of keratinizing colonies from single cells, *Cell* 6, November 1975: 331–343.
28. A. McGehee Harvey, Johns Hopkins—The birthplace of tissue culture: The story of Ross G. Harrison, Warren Y. Lewis, and George O. Gey, *Johns Hopkins Medical Journal Supplement,* 1976: 114–123; also, R. Skloot, *The Immortal Life of Henrietta Lacks*, New York: Random House, 2010.
29. A. Kneller, Stem cell treatment for burn patients earns alpert prize. *Harvard Medical School News,* September 3, 2010; H. Green, *Therapy with Cultured Cells.* Singapore: Pan Stanford Publishing, 2010.
30. J. G. Rheinwald and H. Green, Serial cultivation of strains of human epidermal keratinocytes: The formation of keratinizing colonies from single cells, *Cell* 6, November 1975: 331–343.
31. Howard Green founded a company, Genzyme, that grew skin for burn patients. It was subsequently sold to Biosurface. Two other companies, J-TEC, in Japan, and Tego Science, in South Korea, use Green's method to grow skin. See, A. Kneller. Stem cell treatment for burn patients earns alpert prize. *Harvard Medical School News,* September 3, 2010.
32. See note 14.
33. A comprehensive book about the history of IVF is: H. W. Jones, Jr., *In Vitro Fertilization Comes to America,* Williamsburg, Virginia: Jamestowne Bookworks, 2014. Jones and his wife, Dr. G. Jones, a reproductive endocrinologist, retired from Johns Hopkins University and went down to Norfolk, Virginia, where they opened an in vitro fertility clinic at Eastern Virginia Medical School. The clinic was subsequently named the Jones Institute for Reproductive Medicine. The birth in 1981 of Elizabeth Carr, the first IVF child born in the United States, immediately put both the clinic and the researchers on the front page of the local papers. At first the headlines were unfavorable, as Jones reports in Chapter 7 of his book (pp. 95–100). He even ended up suing the local newspaper, the *Virginian Pilot*, for libel. Eventually, as IVF became a more acceptable procedure, the opposition dissipated. The procedure never gained the approval of the Catholic Church. Jones describes a letter his wife wrote to the pope in 1987 in which she elaborated on the 1968 papal encyclical *Humanae Vitae*. Howard Jones passed away in 2015 at the age of 104.
34. L. Brown and M. Powell, *Louise Brown: My Life as the World's First Test-Tube Baby*, London: Tangent Books, 2015. See also, S. D. James. Test tube baby Louise Brown Turns 35, *ABC News,* July 25, 2013; See also, M. Davies, World's first "test-tube baby" reveals her mother received blood-splattered hate mail when she was born—including a letter containing a plastic foetus. *Daily Mail*, July 24, 2015. A scholarly article on the many changes in IVF technology is, R. M. Kamel, Assisted reproductive technology after the birth of Louise Brown, *Journal of Reproduction and Infertility* 14, July–September 2013: 96–109.
35. Public Broadcasting System. American Experience: Test Tube Babies, October 23, 2006. The DVD includes interview with doctors involved in the IVF, ethicists, and the parents of some of the first children born, as well as interviews with some of the IVF adults. Louise Brown was 25 in July 2003.
36. The papal encyclical called *Humanae Vitae* (Of Human Life), issued in 1968 is still in force today. In that encyclical, Pope Paul VI condemned birth control pills. Only natural conception was accepted. By logical extension the encyclical also prohibited IVF. Today, it also includes opposition to research using human embryonic stem cells. Humanae Vitae: Encyclical of his Holiness Paul VI on the regulation of birth control, July 25, 1968.
37. Two of the major membership organizations try to maintain a payment guideline for both egg and sperm donors. But as the Lewin article clearly illustrates, it is very hard to enforce a guideline. See T. Lewin, Egg donors challenge pay rates, saying they shortchange women, *New York Times*, October 17, 2015. http://www.nytimes.com/2015/10/17/us/egg-donors-challenge-pay-rates-saying-they-shortchange-women.html.

38. H. W. Jones, Jr., *In Vitro Fertilization Comes to America*, Williamsburg, Virginia: Jamestowne Bookworks, 2014.
39. C. C. Miller, Freezing eggs as part of employee benefits: Some women see darker message, *New York Times*, October 14, 2014; also, D. Friedman, Perk up: Facebook and Apple now pay for women to freeze eggs, *NBC News*, October 14, 2014; http://www.nbcnews.com/news/us-news/perk-facebook-apple-now-pay-women-freeze-eggs-n225011.
40. L. Green, Embryo wars: Avoiding Sofia Vergara's legal woes, MSNBC, August 3, 2015; http://www.msnbc.com/msnbc/embryo-wars-sofia-vergara-legal-woes; also, N. Loeb, Sofia Vergara's ex-fiancé: Our frozen embryos have a right to live, *New York Times*, April 30, 2015.
41. Evans, M. J. and M. H. Kaufman. Establishment in culture of pluripotential cells from mouse embryos, *Nature* 292, July 9, 1981: 154–156; also, G. R. Martin, Isolation of pluripotent cell line from early mouse embryos cultured in medium conditioned by teratocarcinoma stem cells, *Proceedings of the National Academy of Sciences* 78, December 1981: 7643–7638.
42. R. M. Green, Report of the human embryo research panel, *Kennedy Institution of Ethics Journal* 5, March 1995: 83–84; R. M. Green et al., The politics of human-embryo research, *New England Journal of Medicine* 335, October 1996: 1243–1244.
43. H. W. Jones, Jr., *In Vitro Fertilization Comes to America*, Williamsburg, Virginia: Jamestowne Bookworks, 2014.
44. The doctor who performed the bone marrow transplant on Molly Nash using her brother's blood was accused of playing God. As the news article notes, the ethical debate has subsided. Today, the reproductive technologies used have become mainstream. See A. M. Faison, The miracle of Molly, *Denver Magazine,* August 2005; http://www.5280.com/magazine/2005/08/miracle-molly?page=full; also, J. Marcotty, 'Savior sibling' raises a decade of life-and-death questions, *Star Tribune*, September 2010; http://www.startribune.com/savior-sibling-raises-a-decade-of-life-and-death-questions/103584799/; also, J. Robertson. Embryo screening for tissue matching, *Fertility and Sterility* 82, August 2004: 290–291.
45. S. Sheldon and S. Wilkinson, Hashmi and Whitaker: An unjustifiable and misguided distinction? *Medical Law Review* 12, 2004: 137–163; also, S. Sheldon and S. Wilkinson, Should selecting savior siblings be banned? *Journal of Medical Ethics* 30, 2004: 533–537.
46. J. Picoult, *My Sister's Keeper*, New York: Atria, 2003.
47. R. M. Green, *Babies by Design: The Ethics of Genetic Choice,* New Haven, CT: Yale University Press, 2007.
48. I. Levin, *The Boys from Brazil*, New York: Random House, 1976. A movie of the book was made in 1978 staring Gregory Peck and Laurence Olivier.
49. M. R. Green and J. Sambrock, *Molecular Cloning: A Laboratory Manual,* Cold Spring Harbor, New York: Cold Spring Harbor Laboratory Press, 2012.
50. K. Takahashi and S. Yamanaka, Induction of pluripotent stem cells from mouse embryonic and adult fibroblast cultures by defined factors, *Cell* 126, August 2006: 663–676.
51. J. B. Gurdon, From nuclear transfer to nuclear reprogramming: The reversal of cell differentiation, *Annual Review of Cell and Developmental Biology* 22, 2006: 1–22; also, R. Williams, Sir J. Gurdon, Godfather of cloning, *Journal of Cell Biology* 181, April 2008: 178–179.
52. C. Mummery et al., *Stem Cells: Scientific Facts and Fiction,* London: Elsevier, 2011. Chapter 6 in the book has a history of how Dolly was created and about her life at the Roslin Institute in Scotland.
53. B. Shapiro, *How to Clone a Mammoth: The Science of De-Extinction*, Princeton, New Jersey: Princeton University Press, 2015: 79.
54. A. Fiester, Ethical issues in animal cloning, *Perspectives in Biology and Medicine* 48, 2005: 328–343.
55. C. Kubota et al., Six cloned calves produced from adult fibroblast cells after long term culture, *Proceeding of the National Academy of Sciences* 97, February 2000: 990–995; also, C. Kubota et al., Serial bull cloning by somatic cell nuclear transfer, *Nature Biotechnology* 22, May 2004: 693–694. Yang and his team set the stage for the cloning of prize livestock. He also realized that this technique could be used in human therapeutic cloning to create cells that were compatible with the patient's own immune system.
56. A. Park, Dogged pursuit, *Time* 166, November 10, 2005: 217.
57. W. Roush, Genetic savings and clone: No pet project, *MIT Technology Review* 108, March 2005: 31.

58. B. Shapiro, *How to Clone a Mammoth: The Science of De-Extinction*, Princeton, NJ: Princeton University Press, 2015.
59. G. Vogel, Breakthrough of the year: Capturing the promise of youth, *Science Magazine* 286, December 1999: 2238–2239. The article notes that since the announcement in 1998 that researchers had succeeded in keeping embryonic and fetal human cells, more than a dozen landmark papers on the "remarkable abilities of these so called stem cells," have been published. The article also commends the work, "which raises hopes of dazzling medical applications" p. 2238.
60. J. A. Thomson et al., Embryonic stem cell lines derived from human blastocysts, *Science* 282, November 1998: 1145–1147; also, M. J. Shamblott et al., Derivation of pluripotent stem cells from cultured human primordial germ cells, *Proceedings of the National Academy of Sciences* 95, November 1998: 13726–13731.
61. G. R. Martin and M. J. Evans, Differentiation of clonal lines of teratocarcinoma cells: Formation of embryoid bodies in vitro, *Proceedings of the National Academy of Sciences* 72, April 1975: 1441–1445; also, G. Martin, Isolation of pluripotent cell lines from early mouse embryos cultured in medium conditioned by teratocarcinoma stem cells, *Proceedings of the National Academy of Sciences* 78, December, 1981: 7634–7638.
62. A. B. Parson, *The Proteus Effect: Stem Cells and Their Promise for Medicine*, Washington, DC: Joseph Henry Press, 2004: 145–147: also, A. Park, *The Stem Cell Hope: How Stem Cell Medicine Can Change Our Lives*, New York: Hudson Street Books, 2011: 66–68.
63. By 1996, the Dickey–Wicker Amendment that prohibited the use of public funds for the creation of human embryos for research purposes, had been passed. Public Law No. 104-99, 1996, called *The Balanced Budget Downpayment Act.*
64. A. Park, *The Stem Cell Hope: How Stem Cell Medicine Can Change Our Lives*, New York: Hudson Street Books, 2011, 78–86.
65. C. Fox, *Cell of Cells: The Global Race to Capture and Control the Stem Cell*, New York: Norton, 2007.
66. Knoepfler lab stem cell blog, *The Niche*, April 6, 2012. Normally we would be suspicious of blogs but Paul Knoepfler, who publishes this one, is one of the leading stem cell researchers in the United States. See P. Knoepfler, *Stem Cells: An Insider's Guide*, New Jersey, World Scientific, 2013.
67. As early as 1994 a research team in Singapore trying to grow hESCs identified the difficulty, noting that it was difficult to get high-quality embryos. A. Bongso et al., Fertilization and early embryology: Isolation and culture of inner cell mass cells from human blastocysts, *Human Reproduction* 9, November 1994: 2110–2117.
68. K. Takahashi and S. Yamanaka, Induction of pluripotent stem cells from mouse embryonic and adult fibroblast cultures by defined factors, *Cell* 126, August 2006: 663–676.
69. J. Yu et al., Induced pluripotent stem cell lines derived from human somatic cells, *Science* 318, December 2007: 1917–1920.
70. E. Lujan et al., Early reprogramming regulators identified by prospective isolation and mass cytometry, *Nature* 521, May 2015: 352–356.
71. T. Vierbuchen et al., Direct conversion of fibroblasts to functional neurons by defined factors, *Nature* 463, January 2010: 1035–1041, L. Qiang et al., Instant neurons: Directed somatic cell reprogramming models of central nervous system disorders, *Biological Psychiatry* 75, June, 2014: 945–951.
72. M. H. Chin et al., Induced Pluripotent stem cells and embryonic stem cells are distinguished by gene expression signatures. *Cell Stem Cell* 5, July 2009: 111–123; also, S. Yamanaka, Induced pluripotent stem cells: Past, present, and future, *Cell Stem Cell* 10, June 14, 2012; 678–684.
73. W. S. Hwang et al., Evidence of a pluripotent human embryonic stem cell line derived from a cloned blastocyst, *Science* 303, March 2004: 1669–1674.
74. W. S. Hwang et al., Patient-specific embryonic stem cells derived from human SCNT blastocysts, *Science* 308, June 2005: 1777–1783.
75. H. Gottweis and R. Triedl, South Korean policy failure and the Hwang debacle, *Nature Biotechnology* 24, February 2006: 114–143; also, G. Kolata, A cloning scandal rocks a pillar of science publishing, *New York Times*, July 7, 2010.
76. H. Obokata et al., Stimulus-triggered fate conversion of somatic cells into pluripotency, *Nature* 505, January 2014: 641–647; H. Obokata et al. Bidirectional development potential in reprogrammed cells with acquired pluripotency, *Nature* 505, January 2014: 676–680.

77. J. Rasko and C. Power, What pushes scientists to lie? The disturbing but familiar story of Haruko Obokata, *The Guardian*, February 18, 2015.
78. U.S. Food and Drug Administration, *Consumer Health Information*, FDA Warns about Stem Cell Claims, January 2012; www.fda.gov/downloads/forconsumers/consumerupdates/ucm286213.pdf.
79. ISSCR, *Patient Handbook on Stem Cell Therapies*, December 3, 2008.
80. NIH, *NIH Clinical Research Trials and You*, November 16, 2015.
81. R. Connolly et al., Stem cell tourism—A web-based analysis of clinical services available to international travelers, *Travel Medicine and Infectious Disease* 12, December 2014: 695–701; also, K.C. Gunter et al., Cell therapy medical tourism: Time for action, *Cytotherapy* 12, 2010: 965–968.
82. *CBS News Canada*, Frustrated former patients question probes into controversial stem-cell researcher, July 2015.
83. D. Cyranoski, Controversial stem-cell company moves treatment out of the United States, *Nature News* 493, January 2013; 12332.
84. S. Juan, Health authority announces step to rein in "wild" stem cell treatment, *China Daily*, August 21, 2015.
85. M. Cook, Scepticism greets new stem-cell regulations in China, *BioEdge*, August 2015. http://www.bioedge.org/bioethics/scepticism-greets-new-stem-cell-regulations-in-china/11547.
86. FDA, *Drug Approval Process;* www.FDA.gov/downloads/drugs/resourcesforyou/consumers/ucm284393.pdf.
87. NIH, *Clinical Trials;* www.nim.hih.gov/services/ctphases.html.
88. Christopher and Dana Reeve Foundation, Paralysis Resource Center: *Facts and Figures.*
89. J. Alper. Geron gets green light for human trial of ES cell-derived product, *Nature Biotechnology* 27, 2009: 213–214.
90. B. J. Cummings et al., Human neural stem cells differentiate and promote locomotor recovery in spinal cord-injured mice, *Proceedings of the National Academy of Sciences* 102, September 2005: 14069–14074.
91. V. Nathan, Asterias's stem cell therapy shows promise in study, *Health*, August 2015.
92. NIH, National Eye Institute, Statistics and Data; https://nei.nih.gov
93. S. D. Schwartz et al., Embryonic stem cell trials for macular degeneration: A preliminary report, *Lancet* 379, February 25, 2012: 713–720.
94. S. Boseley, First UK patient receives stem cell treatment to cure loss of vision, *The Guardian,* September 2015. The research is under the guidance of Drs. Pete Coffey and Lyndon Da Cruz.
95. J. Kyodo, Riken to resume retinal iPS transplant study in cooperation with Kyoto University, *The Japan Times,* June 7, 2016; also, D. Cyranoski, Next-generation stem cells cleared for human trial, *Nature,* September 12, 2014.
96. F. W. Pagliuca et al., Generation of functional human pancreatic B cells in vitro, *Cell* 159, October 9, 2014: 428–439.
97. C. Fox, Embryonic stem cells in trial for diabetes, *Bioscience Technology*, October, 16, 2014; also, A safety, tolerability, and efficacy study of VC-01 combination product in subjects with type I diabetes mellitus, NIH, *Clinical Trials.*
98. J. Millman, Does it really cost $2.6 billion to develop a new drug? The *Washington Post*, November 2014; Tufts University, Tufts Center for the Study of Drug Development, Cost to develop and win market approval for a new drug is $2.6 billion, *News,* November 18, 2014; http://csdd.tufts.edu/news/complete_story/pr_tufts_csdd_2014_cost_study.
99. S. Prakash and V. Valentine, *Timeline: The Rise and Fall of Vioxx*, National Public Radio, November 10, 2007.
100. A. M. Issa, 10 years of personalized medicine: How the incorporation of genomic information is changing practice and policy, *Personalized Medicine* 12, 2014: 1–3.

3 Stem cell federalism

A legal and regulatory quilt

Unlike a unitary nation in which the central government reigns supreme and laws are uniformly applied throughout the country, the United States Constitution created a unique form of power sharing between the states and the federal government. Federalism sought to limit government by dividing power between two levels of government—the federal government (with enumerated powers) and state governments (with open-ended powers). Dual federalism gave the states the right to legislate in areas not assigned to the federal government. This compartmentalization of powers changed over time, especially after World War II, as the federal government gradually preempted state policies in an effort to ensure the equal treatment of all citizens. This new form of shared power has been referred to as cooperative or new federalism.[1] When contradictions arise among state policies, the federal government can either preempt state laws by passing a statute that applies uniformly throughout the nation, or it can let the differences stand until a controversy brings the issue to the attention of the Supreme Court. The Court can then choose to standardize a policy or permit the coexistence of state differences. In many areas of public policy, both levels of government remain active, occasioning contradictory outcomes. Stem cell research is one such area in which federal and state policies have not always provided consistent guidance.

The other area of tension reflected in this policy lies between the president and Congress. The separation of powers originally envisioned in the U.S. Constitution has given way to friction between these two branches as each seeks to promote policy leadership, with the decisive vote sometimes made by the third branch, in the form of the Supreme Court. In this chapter, we first examine federal efforts to regulate this new research area and tensions that emerged between Congress, the president, the National Institutes of Health (NIH), and even the public. The second part of the chapter looks at the divergent role of state governments as some moved to support research, while others sought to ban it.

The federal conundrum

Laws are not immutable. Even the Constitution can be amended, although reinterpretation by a Supreme Court decision is a more frequent avenue for change. An example with some similarity to stem cell policy is that of abortion. This example is also relevant because some commentators equate abortion with hESC research.[2] In 1973, the Supreme Court ruled, in *Roe v. Wade*, that a woman may choose to have an abortion until the fetus becomes viable based on the right to privacy contained in the due process clause of the Fourteenth Amendment. The Court went on to define viability as the "ability of the fetus to live outside the womb, which at that time usually happened between 24 and 28 weeks after conception."[3] As a result, all existing state laws prohibiting abortion were invalidated, though there was no federal law permitting the procedure. The Congress could have passed a federal law banning abortion to overrule the Supreme Court decision, but a majority for such action did not exist in either chamber. Instead, in 1976 Congress passed the Hyde Amendment, barring federal funding for abortions. This rider has been added to all appropriations bills and prohibits the Department of

Health and Human Services, which operates the Medicaid program, from funding abortions.[4] Following *Roe*, some states opted to use their own monies to pay for abortions; other states sought to restrict access to the procedure.[5] This variation in state approaches to policy issues is similarly reflected in the approach to stem cell research.

Two events characterize the legal landscape before the 1998 announcement that hESC had been isolated. The first was the development of IVF and the 1978 birth, in England, of the first baby conceived using this technique.[6] The second occurred in 1981 when Martin and Evans reported the successful isolation of pluripotent stem cell lines from mouse embryos.[7] Both events were hailed by scientists as major breakthroughs, albeit controversial ones. It was not difficult for the astute observer to put the two discoveries together; once you can create a human embryo using IVF, the ability to create hESC was just a matter of time. It actually took almost 20 years, which may be a good benchmark for the time needed to move from lab to clinic. Government panels and commissions were convened in the United States and abroad to discuss the implications of this new discovery and determine if government regulation was needed. Two of the other countries that convened government commissions were Great Britain and Canada. All three groups came up with similar recommendations, and it was on the basis of the recommendations from these three commissions that the Wisconsin Institutional Review Board (IRB) approved James Thomson's protocol for research on human embryos in 1995.[8]

In 1982, the British government established the Committee of Inquiry into Human Fertilisation and Embryology to review developments in these fields. Policy makers were concerned that the speed at which these technologies were developing might require some governmental review, and the role of the committee was to develop principles for the regulation of IVF and embryology. The committee was chaired by Mary Warnock, hence its report is often referred to as the Warnock Report.[9] The committee concluded that the human embryo should be protected, but that research on embryos and IVF should be permissible, given appropriate safeguards. The committee proposed the establishment of a regulatory authority with the responsibility for licensing the use of IVF treatments and storage of and research on human embryos outside the body. This regulatory body, the Human Fertilisation and Embryology Authority, established in 1990, still exists today. The committee's report set out a condition for conducting research on human embryos, stating that "legislation should provide that research may be carried out on any embryo resulting from in vitro fertilisation, whatever its provenance, up to the end of the fourteenth day after fertilisation, but subject to all other restrictions as may be imposed by the licensing body."[10] The end of this 14-day window marks development of what is called the "primitive streak," defined as "the elongated band of cells that forms along the axis of a developing fertilized egg … that is considered a forerunner of the neural tube and nervous system."[11] Basically, one line of argument is that before day 14 all you have is a mass of cells, whereas after 14 days, the cells have started to differentiate and you have the beginnings of the fetal brain.

In Canada, a Royal Commission on New Reproductive Technologies was created in 1989 to study the ethical, social research, and legal implications of the new reproductive technologies. Chaired by Dr. Patricia Baird, the Baird Commission's final report, which was issued in 1993 and codified into law in 2004, permitted the use of human embryos and stem cells in research.[12] This left us with two countries permitting research on human embryos created for IVF.

In the U.S. Congress, there was also a growing concern that IVF research should be revisited. Although a report had been issued in 1979 by an Ethics Advisory Board convened by the Department of Health, Education, and Welfare (now the Department of Health and Human Services), no action had been taken.[13] As we pointed out in Chapter 2, IVF research in the United States was left to the private sector. In 1993, Congress enacted the National Institutes of Health Revitalization Act, providing authority for the NIH to support human embryo research.[14] In response, the NIH created the Human Embryo Research Panel to recommend guidelines for reviewing applications for federal research funds intended to support embryological research. In September 1994, the panel endorsed human embryo research, finding that "the promise of human benefit from research is significant, carrying great potential benefit to infertile couples, families with genetic conditions, and individuals and families in need of effective therapies for a variety of diseases."[15] In making this endorsement, the panel recommended that federal funding be used to support research involving both spare embryos left over from IVF and embryos created specifically for research purposes. This last element may have been a red flag to opponents of stem cell research. To this day, creation of embryos exclusively for research purposes is not permitted. An area found unacceptable by opponents of research on embryos was any "research beyond the onset of closure of the neural tube."[16] If this sounds familiar, it is because the Warnock Commission also used this terminology when it referred to the 14-day window (the time before development of the primitive streak).

While the NIH Director accepted the panel's recommended guidelines, President Bill Clinton issued a statement saying, "I do not believe that federal funds should be used to support the creation of human embryos for research purposes, and I have directed that NIH not allocate any resources for such research."[17] Three factors were likely in play here. First, the 1994 midterm elections had produced a solid Republican majority in the House of Representatives. Second, President Clinton was up for reelection in 1996, and abortion was still a contested issue. In 1992, the Supreme Court had issued a decision in *Planned Parenthood v. Casey* that upheld a state's right to regulate abortion.[18] Third, Congress had passed the Dickey–Wicker Amendment in 1995 as a rider on the annual appropriations for the U.S. Department of Health and Human Services (HHS).[19] The Amendment, named for its authors, Representatives Jay Dickey (R-AR) and Roger Wicker (R-MS), prohibits federal funding for research that involves the creation or destruction of human embryos. It read as follows:

> SEC. 509 (a) None of the funds made available in this Act may be used for:
>
> 1. The creation of a human embryo or embryos for research purposes
> 2. Research in which a human embryo or embryos are destroyed, discarded, or knowingly subjected to risk of injury or death greater than that allowed for research on fetuses in utero under 45 CFR 46.204(b) and section 289(b) of the Public Health Service Act (42 U.S.C. 289(g(b))
>
> (b) For purposes of this section, the phrase "human embryo or embryos" shall include any organism, not protected as a human subject under 45 CFR 46 as of the date of enactment of this Act, that is derived by fertilization, parthenogenesis, cloning, or any other means from one or more human gametes.

The Amendment has been regularly attached to every appropriations legislation since 1996. Given that hESCs had not yet been isolated when the Amendment passed, it was largely a moot point. However, its restrictions came into play during the Clinton, Bush, and Obama presidencies. All three presidents attempted to strike a balance between stem cell research and Dickey–Wicker. There is disagreement about the wording of the Amendment, and if it explicitly or implicitly prohibits stem cell research.

The separation of powers established in the Constitution produced discord between the two branches as each sought authority over the stem cell debate. In this next section, we will observe how Congress, via the Dickey–Wicker Amendment, created a conundrum for the president and his executive agencies.

The Clinton years

At the same time that Congress was debating the Dickey–Wicker Amendment, President Clinton issued Executive Order 12975 (October 1995), creating the National Bioethics

Advisory Commission, with a responsibility to consider issues from human cloning to the use of human biological materials.[20] The Commission was in existence until 2001 and produced a number of reports, including three volumes on ethical issues in human stem cell research.[21] The Commission's report, sent to Clinton in September 1999, affirmed support for stem cell research, noting that research should be limited to cadaveric fetal material and embryos remaining from fertility treatments. It concluded that "recent developments in human stem cell research have raised hopes that new therapies will become available that will serve to relieve human suffering.... [I]n light of public testimony, expert advice, and published writings, we have found substantial agreement among individuals with diverse perspectives that although the human embryo and fetus deserve respect as forms of human life, the scientific and clinical benefits of stem cell research should not be foregone."[22] This need to project a balance between research on hESC and the expected benefits is a recurring theme that we see throughout presidential pronouncements.

Clinton also relied on an opinion provided by the General Council of HHS, Harriet Rabb, as he prepared to approve NIH guidelines for federal funding of hESC. An excerpt from her opinion states that:

> [A] human embryo, as the term is virtually universally understood, has the potential to develop in the normal course of events into a living human being. Pluripotent stem cells do not have the capacity to develop into a human being, even if transferred to a uterus. Therefore, in addition to falling outside of the legal definition provided by statute, pluripotent stem cells cannot be considered human embryos consistent with the commonly accepted or scientific understanding of that term. Thus, based on an analysis of the relevant law and scientific facts, federally funded research that utilizes human pluripotent stem cells would not be prohibited by the HHS appropriations law prohibiting human embryo research, because such stem cells are not human embryos.[23]

Rabb's opinion played a decisive role in the issuance of NIH draft guidelines for hESC funding. She argued that after human embryonic stem cells are derived, or separated from an embryo, the cells no longer constitute an embryo and thus are eligible for federally funded research. By separating the derivation of stem cell lines from research on these lines, the NIH could claim that it was adhering to the requirements of the Dickey–Wicker Amendment. This separation argument became the basis for future controversy and court action. The draft guidelines were issued in December 1999; the NIH began accepting grant proposals in late 2000 when the final guidelines went into effect.[24] No grants were ever awarded and, as the Clinton Administration ended (January 2001) and the Bush Administration assumed control of the White House, the issue of hESC funding remained unresolved. In fact, stem cell funding was one of the first major issues addressed by President George W. Bush.

The Bush years

Almost as soon as he took office, President Bush was confronted by advocates on both sides of the hESC funding issue. In February 2001, 80 Nobel Laureates sent a letter urging the president to continue funding stem cell research. The first paragraph of the letter makes an important point that "research within the rigorous constraints of federal oversight and standards" was "essential to translate this discovery into novel therapies for a range of serious and currently intractable diseases."[25] Federal oversight, along with funding, would ensure research integrity. In retrospect, full funding by the federal government might have curtailed the duplication of state research efforts to fund biomedical research. As we will discuss later in the chapter, a number of states set up their own mini institutes of health to fund research once President Bush limited funding.

On the other side of the issue was the Catholic Church; Pope John Paul II took advantage of a meeting with the president in July 2001 to state his opposition to all research on human embryos.[26] The meeting with the Pope was preceded by a letter from the U.S. Conference of Catholic Bishops calling embryonic stem cell research immoral and unnecessary.[27] Other groups, such as the National Right to Life Committee and the Coalition of Americans for Research Ethics, as well as key members of Congress, also opposed federal funding.

In the end, President Bush took a compromise position. In his first televised speech to the nation on August 9, 2001, he addressed the issue of stem cell research. After setting out the science basics followed by the ethical and moral concerns, he surprised scientists by permitting federal funding on stem cell lines derived prior to his speech. It was clear that the president was conflicted in his decision to fund hESC research. He tried to strike a balance, noting that "scientists believe further research using stem cells offers great promise that could help improve the lives of those who suffer," however, "this issue forces us to confront fundamental questions about the beginnings of life and the ends of science." He did not mince words when he asked the rhetorical question, "Are these embryos human life?" He found ethicists on both sides of the question. In a second rhetorical question, he asked, "If these are going to be destroyed anyway why not use them for good purposes?"[28] Again, Bush noted that there were different responses. In the end he concluded:

> As a result of private research, more than 60 genetically diverse stem cell lines already exist. They were created from embryos that have already been destroyed, and they have the ability to regenerate themselves indefinitely, creating ongoing opportunities for research. I have concluded that we should allow federal funds to be used for research on these existing stem cell lines, where the life-and-death decision has already been made.
>
> Leading scientists tell me research on these 60 lines has great promise that could lead to breakthrough therapies and cures. This allows us to explore the promise and potential of stem cell research without crossing a fundamental moral line by providing taxpayer funding that would sanction or encourage further destruction of human embryos that have at least the potential for life.

President Bush ended his speech by promising to create a council to monitor stem cell research, examine ethical issues surrounding scientific innovation, and recommend ethical research guidelines.[29] The resulting NIH guidelines[30] (issued November 2001) specifically stated that:

- The derivation process (which begins with the destruction of the embryo) was initiated prior to 9:00 P.M. EDT on August 9, 2001.
- The stem cells must have been derived from an embryo that was created for reproductive purposes and was no longer needed.
- Informed consent must have been obtained for the donation of the embryo and that donation must not have involved financial inducements.

The number of stem cell lines eligible for federal funding was unclear. The president said in his speech that more than 60 lines were available. However, the NIH had issued a report in June 2000.[31] The NIH revised that number to 78 eligible lines as of August 9, 2001.[32] Later in the year, when scientists learned more about the lines, they found that while 78 were eligible, only 22 lines were available and that some of these were contaminated by nonhuman animal products. It had been only 2 years since Thomson had derived the first hESC line. These early lines were grown on mouse feeder cells and were considered unsuitable for clinical use, whereas subsequent lines were grown using different cultures and isolation techniques to improve the quality of hESC. Moreover, most of the available lines were derived in foreign countries—England, Australia, India, Israel, and South Korea—and getting the cells to the United States required importation agreements. The NIH provided information on how to secure the agreements, but scientists were not enthusiastic about using these lines in their research. The lines available were derived from embryos left over from IVF treatments and carried few genetic diseases, whereas scientists were interested in studying lines with different genetic conditions so they could study disease progression.

The NIH issued guidelines following President Bush's August 9, 2001 speech.[33] While the word "guidelines" may sound advisory or even voluntary, these were not advisory. Contrary to some reports, the president did not issue an Executive Order in 2001 (he did so in 2007). Rather, he directed the NIH to issue guidelines consistent with his policy directives as set out in his August 9, 2001 address. Agencies such as the NIH get their authority to issue regulations from statutes (laws) enacted by Congress, as well as presidential authority delegated to the agency. President Bush, under his authority as head of the executive branch, directed the NIH to issue guidelines. That these guidelines had the force of law can be seen by the fact that the NIH followed the requirements set out in the Administrative Procedures Act (APA).[34] Once the guidelines were developed, they were published in the Federal Register as draft guidelines, and the NIH solicited comments from the public. After considering the public feedback, the NIH made changes and then published the final guidelines in the Federal Register, specifying the effective date. The Federal Register is the official daily publication for all agency rules, proposed rules, as well as executive orders and other presidential documents. The APA also lays out the process for judicial review of rules in federal court. For example, individuals and corporations may go into court to make a claim that they have been or will be damaged or adversely affected in some manner by regulations. This happened in 2010 when two researchers sued the NIH under the Obama Administration's guidelines. This lawsuit is discussed later in the chapter.

The NIH issued a number of guidelines over the course of the Bush Administration, and the texts of these guidelines can be found in the NIH policy and guidelines archive. The eligibility of federal funding for hESC lines, however, continued throughout the Bush Administration to be limited to those derived prior to August 2001. It is this date, which remained fixed until the Obama Administration, that was the cause of distress for scientists. Over the course of the Bush years, a number of new hESC lines had been derived and were being offered to interested researchers, but only scientists with non-federal funding could work with these lines.[35]

On June 22, 2007, President Bush issued E.O. 13435: Expanding Approved Stem Cell Lines in Ethically Responsible Ways. The E.O. states that, "within 90 days of this order, the Secretary … shall issue a plan, including such mechanisms as requests for proposals, requests for applications, program announcements and other appropriate means, to implement subsection (a) of this section…."[36] Three of the president's directives were:

1. Prioritize research with the greatest potential for clinical benefits.
2. Rename the "Human Embryonic Stem Cell Registry" the "Human Pluripotent Stem Cell Registry."
3. Add to the registry new human pluripotent stem cell lines that are derived without creating a human embryo.

The intent of the order was to encourage scientists to use the newly discovered induced pluripotent stem cells (iPSC) in research. The implementation plan developed by the NIH and published in September 2007 was an attempt at outreach and dissemination rather than a new research agenda.[37] The six action items included:

1. Issue Funding Opportunity Announcements
2. Rename the NIH Stem Cell Registry
3. Undertake a Comprehensive Portfolio

4. Consider Alternative Sources of Pluripotent Stem Cells
5. Convene a State-of-the-Science Workshop
6. Hold a Human Pluripotent Stem Cell Research Symposium

Although a lot was going on around the globe with stem cell research, the NIH was not viewed as a key player. Privately funded foundations and academic institutes, both in the United States and abroad, had emerged as major players in this new field. As of 2007 no hESC therapies were in clinical trials in the U.S. Research was ongoing, but it was basic research. Renaming the registry was a politically expedient way to get a loaded term, "embryonic," out of the title, but the rest of the E.O. focused on giving the NIH a more prominent role at the research table. By 2007, the science policy literature was filled with articles about the missed opportunities resulting from the Bush Administration's decision to limit funding on stem cell lines.[38]

Battling Congress and patient advocates

President Bush's decision left few happy. Like President Clinton, he also faced a potential legal challenge on the basis that federal funding for hESC would be a violation of the Dickey–Wicker Amendment. One of President Bush's advisors later wrote that the policy question was not whether the research should be made legal; "[n]o law in the country banned it, nor was anyone in either party pressing for such a ban. Rather, the question being put to him was whether he would authorize the use of federal funds."[39] An evaluation of the Dickey–Wicker Amendment by Bush's legal team concluded "that while spending federal dollars on such research might violate the spirit of the amendment, it would not violate the letter."[40] After all, the Amendment prohibits federal funding for "research in which the human embryo or embryos are destroyed,"[41] whereas the Bush policy was limited to funding research on a discrete set of stem cell lines, the embryos in which had been derived (destroyed) using private funding. This same argument will be made during the Obama Administration.

Although Bush enjoyed a Republican majority in both houses of Congress for most of his tenure, this did not guarantee a tranquil coexistence. Beginning in 2004 a bipartisan coalition of House and Senate members attempted to pass legislation expanding the president's hESC funding policy to extend the date for eligible lines past August 2001. The Stem Cell Research Enhancement Act was introduced three times during the Bush Administration and once during the Obama Administration. While all attempts failed, two of the efforts created a critical mass of support, including one attempt to override a presidential veto. Presidential vetoes make news, as do Congressional attempts to override them. President Bush's veto would be the first of his presidency.[42]

All of the bills had the common goal of removing the August 9, 2001 date for derivation of stem cell lines. The three bills were essentially similar: H.R.4682: Stem Cell Research Enhancement Act of 2004 (108th Congress) was sponsored by Mike Castle (D-DE) and had 190 cosponsors. This bill died in committee. However, the other two bills that engendered far more publicity were:

H.R.810	Stem Cell Research Enhancement Act of 2005 (109th Congress)[43]
S.471	Mike Castle (D-DE); Arlen Specter (R-PA) Bill passed H.R. and S. Vetoed by the president. House attempt to override veto, failed.
H.R.3	Stem Cell Research Enhancement Act of 2007 (110th Congress)[44]
S.5	In an effort to force the president to expand stem cell research Diana DeGette (D-DE); Harry Reid (D-NV). Passed H.R and S. Vetoed by the president. No override attempt.

The 2005 Act created significant debate and occasioned a split in the Republican Party between social conservatives, who sided with the president, and more moderate members. Senator Specter (R-PA) represented the more moderate wing. His committee held hearings on the bill and invited top scientists to share their research findings and prospects for future therapies with members of Congress. These hearings had the expected effect of drawing public attention to the issue and also getting some members to drop their opposition. Senate Majority Leader Bill Frist (R-TN) abandoned his opposition to the bill and signaled that he would support expanding the number of stem cell lines available for funding. Frist, a thoracic surgeon, acknowledged that the number of lines eligible for federal funding was 22 rather than the 60 identified in the president's speech. Frist also admitted that, while he was pro-life, he disagreed with the president's decision to veto the Act.[45] Members who voted to override the veto argued that the August 9, 2001 date was arbitrary. Extending the date for derivation of hESC lines would provide researchers with better quality lines, and the derivation would continue to be paid for by private and philanthropic sources. The House attempted to override the president's veto but fell short of the two-thirds majority needed. The Senate did not vote to override.

In vetoing the Act, the president reiterated his views that the bill would disrupt his Administration's balanced policy on hESC and would compel taxpayers to fund the deliberate destruction of human embryos. President Bush commented that the "bill would support the taking of innocent human lives in the hope of finding medical benefits for others."[46] After vetoing the Act, the president attended a White House reception for families whose children were born from adopted frozen embryos left over from IVF treatment. These so-called "snowflake babies" were provided by Nightlight Christian Adoptions, a California-based adoption agency that pioneered the Snowflakes Embryo Adoption Program.[47] Holding a child in his arms, President Bush remarked that "[t]hese boys and girls are not spare parts. They remind us that we all begin our lives as a small collection of cells."[48]

The president made no secret of the fact that he was looking for alternatives to hESC. In his August 9, 2001 address he confirmed that "great scientific progress can be made through aggressive federal funding of research on umbilical cord, placenta, and adult and animal stem cells."[49] So it was in keeping with his promise to support this research when he signed two stem cell-related measures into law during his Administration. The first was the Stem Cell Therapeutic and Research Act of 2005, the second was the Fetus Farming Prohibition Act of 2006. This second Act was passed unanimously in both the House and the Senate, given that there was no evidence that anyone had tried to use tissue that came from an embryo implanted in a woman for research purposes.[50]

The Stem Cell Therapeutic and Research Act, H.R.2520, was introduced by Chris Smith (R-NJ) and Orin Hatch (R-UT) and authorized the "Secretary of HHS to contract with qualified cord blood banks to assist in the collection and maintenance of 150,000 new units of high quality cord blood to be made available for transplantation."[51] The emphasis throughout the Bush Administration was to limit research on hESC in favor of alternatives that used adult/somatic stem cells. Supporters of Bush's approach were quick to point out that there were already many successful therapies using adult stem cells and none using hESC.

Congress tried again to introduce legislation to expand hESC lines eligible for funding in the 110th Congress. The bills, S.5 introduced in the Senate by Harry Reid (D-NV), and H.R.3 in the House, introduced by Diana DeGette (D-DE), were passed in both chambers by a strong bipartisan majority. The president again vetoed the measure. This time neither body attempted an override. To emphasize his resolve, President Bush issued E.O. 13435 on the same day that he vetoed the Congressional measure—June 20, 2007. The E.O. reiterated what the president had earlier established, "that any human pluripotent stem cell lines produced in ways that do not create, destroy, or harm human embryos will be eligible for federal funding."[52]

Between 2006 and 2007, researchers in Japan and the United States derived iPSC by reprogramming skin cells using four specific transcription genes, but the experiments were carried out on mouse cells. This did not deter the president from praising the findings as proof that "medical problems can be solved without compromising either the high aims of science or the sanctity of human life."[53] Bush's E.O 13435 was a way to push scientists to work with human iPSCs rather than hESCs, but it would take more than renaming the NIH Registry to move research in this direction.

Patient advocacy groups

In addition to pushback from Congress, President Bush was lobbied by a number of patient advocacy groups that emerged during his presidency. In part because of the media attention paid to the potential benefits of stem cell research, groups wanted the president to take a more proactive stand toward moving research into the clinic. The more visible groups included the Christopher and Dana Reeve Foundation, the Michael J. Fox Foundation, the Roman Reed Foundation, and the Juvenile Diabetes Research Foundation. Individual advocates also attempted to persuade the president to expand funding for stem cell research. Among the most prominent of these was Nancy Reagan. Not only did President Bush acknowledge her in his August 2001 speech, but she made her position clear in speeches and letters, often mentioning her husband, who was suffering from Alzheimer's. President Ronald Reagan passed away in June 2004. At the Democratic National Convention that year, Ron Reagan, the president's son, gave a speech voicing his support for an expansion of stem cell research.[54]

Christopher Reeve, until his death in October 2004, was the face of stem cell research. A popular actor who suffered a serious spinal cord injury, he became a regular visitor to Capitol Hill, advocating for more stem cell research funding. During the 2004 presidential election, stem cell research became a campaign issue between President Bush and the Democratic challenger, John Kerry, and Reeve did not hesitate to criticize President Bush for his limited support for stem cell research. The Christopher and Dana Reeve Foundation, located in New Jersey, continues to operate and to fund biomedical research with an emphasis on cures for and treatment of individuals with spinal cord injuries.[55]

The Michael J. Fox Foundation for Parkinson's Research was established in 2000 with the goal of finding a cure for Parkinson's disease. Fox, a popular television and film actor, had been an active spokesperson and fundraiser supporting basic research. During the 2006 Congressional elections, Fox made an ad endorsing Missouri Senate candidate Claire McCaskill, (D-MO), due to her support for hESC research. Fox was accused by conservative radio commentator Rush Limbaugh of exaggerating the effects of his disease (Fox's hand tremors). What ensued was a verbal confrontation between the two media personalities. The clash produced a great deal of publicity for the Foundation and for the cause of Parkinson's.[56]

Roman Reed suffered a spinal cord injury while playing college football in California. With the help of his father, he founded the Roman Reed Foundation to support the search for a cure for paralysis. In 2009, he was invited to the White House by President Obama to celebrate the president's reversal of the federal ban on hHEC funding. Foundation support led to the creation of the Roman Reed Core Research Facility at the University of California, Irvine.[57]

The Coalition for the Advancement of Medical Research (CAMR) was founded in 2001 as a peak association representing patient organizations, universities, scientific societies, and foundations. Its mission was bipartisan advocacy for stem cell research with an emphasis on federal funding for hESC research. In 2013, CAMR transferred its mission and assets to the Alliance for Regenerative Medicine (ARM), which is also a peak association. However, ARM's membership is largely comprised of biotechnology and pharmaceutical companies, including Pfizer, Asterias Biotherapeutics, and Novartis.[58] The new emphasis is on product development and investment. To get a sense of how much things have changed, ARM hosts an annual Cell and Gene Therapy Investor Day.

The stem cell field, in its infancy when George Bush became president, had matured and expanded by the time he left office in 2009. Some argued that his restrictive policy on hESC research had been a major setback for biomedical research in the United States. Others pointed out that the United States had maintained its dominance in the field as a result of increased funding from states and the private sector. The Obama years would prove to be less controversial.

The Obama years

Barack Obama campaigned on the promise of reversing the ban on hESC. He kept his promise on March 9, 2009, by issuing the first executive order of his Administration, E.O. 13505, that reversed the Bush limit on hESC lines eligible for federal funding.[59] The NIH was instructed to issue revised Guidelines on Stem Cell Research, which went into effect in July 2009.[60] However, the guidelines added some extra conditions for eligibility, including a lengthy approval process to ensure that:

1. The lines were derived from leftover IVF embryos.
2. Donors were informed in advance that:
   a. Embryos would be used to derive stem cells.
   b. Embryos might be kept indefinitely.
   c. Donation would be made without restrictions on who could benefit.
   d. Research would not provide donor with direct benefits.
   e. Donors would not receive financial benefits from commercial developments.
   f. Donors would be notified if identifiable information would be available to researchers.

The guidelines also stipulated that there had to be a clear separation between the IVF facility and the facility engaged in hESC research.[61] This was to avoid the issue that occurred in South Korea in which egg donors were also involved in research.

Despite the long process of securing informed consent, the number of hESC lines available for federal funding increased initially to over 100 and, by 2015, to over 300.[62] Before scientists could start working with the lines, however, a legal challenge had to be resolved.

The case of *Sherley v. Sebelius*[63] was a lawsuit brought by two adult stem cell researchers, the Nightlight Christian Adoption Agency and the Christian Medical Association, in which the plaintiffs argued that NIH guidelines were in violation of the Dickey–Wicker Amendment. Does this sound familiar? Remember that the NIH does not fund the derivation of hESC lines, only research after derivation.

Initially, Judge Lamberth of the U.S. District Court for the District of Columbia dismissed the case, arguing that the plaintiffs lacked standing because they were not materially (financially) harmed. James Sherley and Theresa Deisher both worked on adult stem cell lines; they pointed out that if more hESC lines were eligible for federal funding, money would be diverted away from adult stem cell research and toward hESC research. This argument was convincing; the U.S. Court of Appeals for the District of Columbia Circuit agreed and sent the case back to the District Court. This time Judge Lamberth ruled in favor of the plaintiffs, noting that on closer reading it did appear that the Obama E.O. was in conflict with the Dickey–Wicker Amendment that prohibits federal funding of "research in which a human embryo or embryos are destroyed, discarded, or knowingly subjected to risk of injury of death."[64]

Judge Lamberth rejected the idea that the research could be separated into derivation of the hESC lines and research on those lines. You might recall that this separation has been continually supported by the NIH. Even the Bush Administration guidelines allowed funding of research on hESC lines derived before his August 2001 speech. The Obama Administration appealed and won a decision by the Court of Appeals for the District of Columbia which agreed that:

> Dickey–Wicker is ambiguous and the NIH seems reasonably to have concluded that, although Dickey–Wicker bars funding for the destructive act of deriving an ESC from an embryo, it does not prohibit funding a research project in which an ESC will be used.[65]

In January 2013, the U.S. Supreme Court declined to hear the case, ensuring the Appeals Court ruling would remain in effect. The news was welcomed on two fronts. In the United States, hESC scientists had many more lines with which to work. But also, across the globe, international researchers collaborating with U.S. universities and centers also applauded the decision that ensured continuation of their joint efforts.[66]

At the time, Congress was less concerned about President Obama's stem cell policy than they were about his Affordable Care Act. Nevertheless, there were four different attempts in the House and one in the Senate to pass the Stem Cell Research Advancement Act.[67] This Act, cosponsored in the House by Rep. Diana DeGette (D-CO) and Rep. Charlie Dent (R-PA) and in the Senate by Arlen Specter (R-PA), was an attempt to codify the NIH guidelines for carrying out stem cell research and required the NIH to review its guidelines every 3 years and make updates as warranted by the science. Basically, there was a fear that once President Obama left office, his E.O. might be revoked. There was precedent for this; in fact, President Obama had revoked the Bush-era E.O. The new bills also included a provision prohibiting any funding for human cloning. Putting the guidelines in statutory form would guarantee that they could only be altered by an act of Congress. The Stem Cell Research Advancement Acts (2009, 2010, 2011, and 2013) are sometimes confused with the Stem Cell Research Enhancement Acts proposed during the Bush Administration. The important difference is that the Enhancement Acts would have changed Bush's policy, while the Advancement Acts would have codified Obama's Executive Order. Since Congress has a solid Republican majority in both houses, it is not likely to pass the Advancement Act before President Obama leaves office.

Despite the economic downturn during the first years of the Obama Administration, stem cell research advanced. State efforts initiated during the Bush years continued to grow. The California Institute for Regenerative Medicine (CIRM) continued to accumulate world-class researchers, as did other state efforts. While human clinical trials were still limited, the few that did exist were beginning to show promise. International collaboration was also in evidence as clinical trials took place in different countries. Researchers were using stem cells for drug testing and to model diseases. With the limits on federal funding largely removed, the controversy between scientists and the executive branch diminished.

In the next section of this chapter, we look at what was going on at the state level that helped advance hESC research.

State initiatives

In The Federalist Papers Number 17, Alexander Hamilton noted that "it will always be far more easy for the State government to encroach upon the national authorities than for the national government to encroach upon the State authorities."[68] Few people, and probably no state administrators, would agree with this statement today. Through a variety of mechanisms—grants-in-aid, preemption, and mandates—the national government has moved into policy areas formerly considered the domain of state governments or altogether outside the realm of government. There is one exception—stem cell research.

As laid out in the first part of this chapter, the federal government, reluctant to move into this area, passed no laws banning research, even cloning is not prohibited (despite efforts by Congress to ban the procedure).[69] The federal prohibition was solely on the use of federal funds to derive hESC. Federal funds could be spent on adult stem cell research and, eventually, hESC and, later, iPSC research. As we look at how the states handled the stem cell research question, it is appropriate to quote another well-known American, Justice Louis Brandeis. In his dissent in *New State Ice Company v. Liebmann*, he remarked that "[i]t is one of the happy incidents of the federal system that a single courageous state may, if its citizens choose, serve as a laboratory; and try novel social and economic experiments without risk to the rest of the country."[70]

President Bush's limit on federal funding for hESC research did not deter some states from engaging in the kind of experimentation to which Brandeis referred. However, Brandeis could not have imagined that in the twenty-first century this experimentation would lead to competition among states. Wealthy states with robust biotechnology sectors took advantage of the federal funding hiatus to pump state money into this emerging field.

In Table 3.1, we provide an overview of the seven states that passed legislation and provided public funding for hESC research. Except where noted, the table does not include private or foundation funding. Some research facilities receive much of their funding from philanthropic donors or private investors, but the exact amounts are difficult to determine. In 1998, the Geron Corporation funded the research of both Thomson and Gearhart.

Once the long-term implications of President Bush's funding limitations were fully calculated, stakeholders (scientists, entrepreneurs, state officials, politicians) sprang into action. States with high per-capita incomes were in a better position to pursue public funding, as were states with thriving biomedical and/or pharmaceutical sectors.[71] Stakeholders feared that the United States (and their individual state) might lose out in the race to become the key stem cell hub. By 2004, when South Korean scientists announced (falsely) that they had cloned a human embryo, there was a palpable fear of the United States losing its biomedical preeminence.[72]

California was the first state to take up the hESC research challenge. Its approach to supporting stem cell research became a model for other states, though none were successful at garnering California's impressive financial support. California is a wealthy state with a large biomedical sector, prestigious universities, and world-class scientists. In 2002, it was the first state to pass a law permitting hESC research. The governor, Grey Davis, had no problem getting the bill through the legislature, but no funds were appropriated.[73] This caused some to point out that the effort was largely symbolic, since researchers were still dependent on federal funding or private grants.

California is one of 26 initiative states, meaning that citizens can petition to put a measure directly before the voters.[74] In November 2004, Proposition 71, the California Stem Cell Research and Cures Initiative, was approved by voters 59–41%. Prop. 71, as it became known, authorized the sale of $3 billion in general obligation bonds over a 10-year period to support stem cell research. Although it was understood that funding was mostly for hESC research, money could also be used for other types of stem cell research. One hundred million dollars per year for 10 years is a substantial amount of money and clearly positioned California to aim for leadership in this growing field. How did proponents convince citizens to fund such a large investment in, as yet, an unproven field?

Proponents of the proposition, led by Robert Klein II, a Palo Alto real estate developer, raised over $35 million from donors to mount a formidable campaign for passage. The money was used to canvas voters and air television ads featuring Hollywood celebrities including Christopher Reeve and Michael J. Fox. A number of events in 2004 lined up to favor the proponents—President Reagan passed away; his widow, Nancy Reagan, and son, Ron Reagan, both vocally supported stem cell research; Democratic presidential candidate John Kerry favored funding. Finally, Arnold Schwarzenegger, the newly elected Republican Governor, came out in support of the measure.

Prop. 71 created the California Institute for Regenerative Medicine (CIRM), modeled after the NIH, which became the initiative's administrative and funding arm. However, before bonds could be sold or funds distributed, CIRM was sued by opponents, including the California Family Bioethics Council, claiming that the law created a taxpayer-funded body that had limited regulatory oversight.[75] The suit was resolved after several years, but the hiatus added to research uncertainty and delay. The initiative received a $150 million loan from the state's general fund that enabled CIRM to begin operating. What did California get as a result of its funding?

Prop. 71 promised voters medical cures, but other selling points were economic benefits. California would become a magnet for biomedical and technology companies, and high-paying jobs would be created with employees adding to the state's tax base. Of course, the intangible

Table 3.1 States permitting and funding hESC research 2001–2009[a]

State	*Statute*	*Agency*	*Public funding*	*Other*
California	125300 320 (2002) Prop. 71: Const. Art. XXXV	California Institute for Regenerative Medicine (CIRM)	No funding $3 billion over 10 years 2007–2017	
New Jersey	Title 26: 2Z-2; 2C; P.L. 2006, Ch. 102	Stem Cell Institute of New Jersey	2004–10 mil; 2005–5 mil; 2006–750 mil (construction)	2007 failed to pass Q2–450 million/10 years[b]
Connecticut	P.L. 05–149 (19a–32 g)	Connecticut Stem Cell Research Committee 2014–Connecticut Innovation	100 million/10 years (2006–2013)	
Massachusetts	Pt. 1. Title XVI, 111L	Department of Public Health–Biomedical Research Advisory Committee Massachusetts Life Science Center	9 million (2008–2013) Mass Life Science Initiative (1 billion/10 years)	Stem Cell Bank & Registry closed 2012 Could fund stem cell research
New York	Art. 2, Title 5-A, (265–265f)	Empire State Stem Cell Board New York Stem Cell Foundation	108 million 36 mil FY16 (suspended) 150 million since/2005	Nonprofit has own lab/funds research other institutions
Maryland	Maryland Stem Cell Research Act 2006; Article 83A, Title 5–2B–01–013	Maryland Stem Cell Research Commission & Maryland Technology Development Corporation (TEDCO)	$129.5 million FY 2006–2016	
Illinois	E.O. 6 (2005)/E.O. 3 (2006) Pub. L. 095–0519 (2007)	Illinois Regenerative Medicine Institute (IRMI)	10 million/5 million 100 million–5 years (2010–2015)	Executive Order

Source: Adapted and updated from the National Conference of State Legislatures, Embryonic and Fetal Research Laws, 2015.

Notes

a 2001–2009, Bush limitations on federal funding, states that used state funds to promote research on hESC. After March 2009, more lines were eligible for federal funding reducing the necessity of states to support research. It is also the beginning of an economic downturn that caused some states to reduce budget obligations.

b Modeled after California Prop. 71, failed 53–47%.

benefit would be the status of California as the leader in domestic stem cell research. An economic impact study commissioned by CIRM in 2011 projected significant economic benefits directly and indirectly from the state's investment in stem cell research.[76]

A recent study (published in 2014) confirms that, based on publications in scholarly journals, California scientists published 43% more articles in the stem cell field compared with published research in other biomedical fields, such as cancer.[77] While publications are just one indicator of California's status in the field, it suggests that research funding produced quantifiable outputs—peer-reviewed articles. Other indicators include 10 clinical trials funded in part by CIRM money,[78] and the fact that CIRM's funding arm has put an emphasis on supporting research that is within 5 years of translation.[79]

The downside of the California experiment is that a lot of money was spent on constructing research facilities that created jobs, at least temporarily. The need to physically use separate facilities when working on the derivation of hESC was dictated by the Bush-era funding policy. Initial CIRM expenditures were heavily weighted toward construction of facilities rather than research. However, the 10-year funding commitment produced the anticipated effect of putting California on the map as a major player in stem cell research. An article published in the *New York Times* in 2005 sums up the growing attraction: "[f]or many years, human embryonic stem cell research was an unwanted stepchild. Conservative evangelical groups opposed it; Congress gave it a mixed report. Presidents shunned it or supported it halfheartedly. But suddenly a glass slipper seems to be fitting this Cinderella, and California may be just the first in a line of rich suitors."[80]

New Jersey was the second state to pass legislation permitting hESC research. A wealthy state, New Jersey is home to 17 of the world's largest pharmaceutical companies—Johnson & Johnson, Merck, and Bristol–Myers Squibb among them—so it is not surprising that the state was eager to become a center for stem cell research. The state awarded $5 million in research grants in 2005 and another 1.5 million for the construction of the Stem Cell Institute of New Jersey, located in New Brunswick, that included the Christopher Reeve Pavilion. The structure was completed, but operating funds sufficient to continue long-range research never materialized.

In 2007, Governor Jon Corzine (D), tried to duplicate California's success by promoting Public Question 2, an initiative that would have authorized $450 million in public obligation bonds over a 10-year period to both construct additional facilities and support stem cell research. The bond measure included funds for the construction of four additional stem cell facilities throughout the state; this was an effort to spread the funding and jobs throughout the state, rather than concentrating them in New Brunswick. Like California, New Jersey also touted job creation as a benefit in addition to the possibility of becoming a magnet for biotechnology companies.[81]

Unlike their California counterparts, the proponents were not well funded. Despite the efforts of New Jersey's top stem cell scientist, Wise Young, and the advocacy of the Christopher Reeve Foundation, the initiative was defeated. Strong opposition from the New Jersey Right to Life organization convinced voters that the measure was both unethical and too expensive.

The defeat was attributed to low voter turnout (2007 was not a presidential election year) and taxpayer fatigue. New Jersey had a budget deficit of over $3 billion at the time, and voters on both sides of the stem cell issue felt that the state should deal with the deficit before authorizing additional long-term spending.[82]

Connecticut was the third state to pass legislation, in 2006, to authorize the use of public funds to finance stem cell research. The law committed $100 million over a period of 10 years, from 2007 to 2017, to support research.[83] Originally, the governor, Jodi Rell (D), wanted a $20 million, 2-year program, but she was convinced by science advisors that the top researchers would not relocate based on a short-term funding commitment. The first 2 years of funding came from general revenues, and the remaining 8 years were dependent on money from the Tobacco Settlement Fund.

When asked about the sum Connecticut would spend on research compared to California, the head of a consortium of Connecticut biotechnology companies, Paul Pescatello, stated, "Their $300 million is spread over probably over a hundred institutions in California. The Connecticut dollars, the $10 million a year, is spread among really two or three or four institutions—I mean mostly University of Connecticut and Yale University and Wesleyan University."[84]

Massachusetts followed Connecticut in trying to position itself as a major player in the hESC field. A number of private institutions had already set up separate labs for conducting hESC research, including the Harvard Stem Cell Institute (HSCI), founded in 2004; The Whitehead Institute for Biomedical Research at MIT, founded in 1982; and Tufts University School of Medicine. Other facilities, such as Boston Children's Hospital and the Dana–Farber Cancer Institute, were also working on stem cell research. Basically, Massachusetts could tap into a lot of private money to fund hESC research. The HSCI, cofounded by Douglas Melton and David Scadden, became one of the country's premier labs for researchers interested in deriving hESC. In 2004, using private funds, Melton and his colleagues created 17 hESC lines and distributed them to researchers around the world without charge.[85] Melton's philosophy was that if others made important discoveries, all the better.[86] This was in contrast to the University of Wisconsin Alumni Research Foundation (WARF), which charged researchers $5000 per vile of hESC. (WARF held patents on the hESC lines developed by James Thomson.)

In 2008, Governor Deval Patrick (D-MA) signed a 10-year $1 billion Life Sciences Initiative that would provide funds for hospitals, universities, and businesses to work together to promote the state as a national hub for medicine and science. The state authorized funds for the creation and operation of the University of Massachusetts Human Stem Cell Bank and Registry, which opened in early 2011 with seven hESC lines developed in the laboratory of George Daley at Children's Hospital. The idea was that hESC lines derived with private funds would be maintained and made available for distribution to researchers worldwide; this would gather all the lines in one location and relieve individual institutions of the expense of cataloging the lines and providing the appropriate paperwork before sending them to other research labs. However, due to financial issues, the Bank closed in 2012. Although money ran out for the bank, the state continues to fund the Massachusetts Life Sciences Center, which can fund stem cell research. Most of the funding to date has gone for capital investments, loans to companies, job creation, and research.

New York State was also an early public supporter of hESC research. The state created the Empire State Stem Cell Board in 2007 to support stem cell research; it supports infrastructure, training, and educational incentives (getting more high school students into STEM programs), in addition to direct research. Since 2007, the Board has awarded a total of $354 million to 323 proposals. In fiscal year 2016, 36 million dollars in grants were suspended due to budgetary issues.

New York is unique in that, in 2005, a nonprofit foundation was created to support stem cell research, the New York Stem Cell Foundation (NYSCF). The Foundation gets money from a number of sources and supports research through a grant-in-aid program. It is not limited to supporting only researchers located in New York. Since 2011, the Foundation has awarded grants to stem cell scientists making significant breakthroughs in translation research.[87] One of the first recipients was Dr. Pete Coffey, whose research using hESCs to cure blindness is currently in clinical trials in England and the United States. Another recipient was Dr. Kazutoshi Takahashi, who worked with Shinya Yamanaka to discover iPSCs.

This foundation has spent more than $150 million since 2005. The Foundation supports research in its own lab in New York City as well as collaborating with labs across the state and the country. It has a significant presence in the field due in part to its cofounder and CEO, Susan Solomon, an attorney and entrepreneur, who spent most of her career building businesses. Maybe that is what is needed to push stem cell research to the next level.

Maryland passed the Stem Cell Research Act in 2006 to promote state-funded human stem cell research and medical treatments. Grants of between $110,000 and $750,000 are administered through the Maryland Technology Development Corporation (TEDCO). This Corporation was created by the state legislature in 1998 to encourage the commercialization of technology from Maryland research universities and federal labs (the NIH and a number of other federal labs are located in Maryland) to the market.[88]

To date, $129 million have been distributed through fiscal year 2016. Although Maryland is a small state, it is home to The Johns Hopkins University and the University of Maryland, as well as a number of biotechnology companies that are eligible for state funding. Johns Hopkins University is one of the largest recipients of federal grants in the science and medical fields. TEDCO tries to leverage funding by encouraging matching grants by state-owned companies.

Illinois began funding hESC research as a result of two executive orders issued by Governor Rod Blagojevich (D-IL) (E.O. 6 [2005] and E.O. 3 [2006]), when he was rebuffed by the state legislature.[89] His first order created the Illinois Regenerative Medicine Institute (IRMI) and allocated $10 million in research funding. A second E.O. added another $5 million to continue funding. In 2007, the state legislature passed the Regenerative Medicine Act, permitting IRMI to conduct research on cells from any source, though funds were not allocated until 2009. Subsequent state legislature has allocated $100 million over a 5-year period (2010–2015).

Wisconsin did *not* provide any state funding for hESC research. Although Dr. James Thomson, the scientist who first isolated hESCs, is a faculty member at the University of Wisconsin–Madison campus, no public funding was used in the isolation of or subsequent research on hESC.

Two weeks after the passage of Prop. 71 in California, Governor Jim Doyle (D-WI) announced plans to invest $750 million to support hESC, but the money was never appropriated by the state legislature. In 2010, when Governor Scott Walker (R-WI) took office, he signaled that he thought that research on adult stem cells was more promising.

A primary source of funding for hESC research is the Wisconsin Alumni Research Foundation (WARF), which is a private, nonprofit organization. In 2002, the WiCell Research Institute, a WARF subsidiary, became the repository for hESC lines (the so-called presidential lines) eligible for federal funding. WiCell is also the home of the National Stem Cell Bank that is funded by the NIH.[90] Massachusetts also tried to support a Stem Cell Bank that included hESC lines derived after August 9, 2001. So in some respects, while Wisconsin did not publicly fund hESC research, through the WiCell Institute the state became the beneficiary of federal funding that could be used to support stem cell scientists.

As an interesting aside, in 2008 when James Thomson was interviewed for a question-and-answer session that was published in a stem cell journal, he was asked whether his recent affiliation with the University of California, Santa Barbara (where he holds a professorship) had anything to do with CIRM funding. He stated, "Clearly part of the interest for me based on this administration is the CIRM funding, but that's not my primary motivation. They have a marine center that's right on the ocean."[91] CIRM's funding is limited to research groups in which at least one member is affiliated with a California institution. Stem cell researchers, especially if well-known, are very mobile and in demand, and there are many opportunities to work in locations other than their home campus.

Summary

Federal funding for hESC was not available until 2001, when President Bush authorized the NIH to issue guidelines permitting funding on lines derived before August 9, 2001. In 2007, he issued an Executive Order reaffirming and expanding on the stipulations set forth in the earlier guidelines. President Obama's Executive Order revoked Bush's Order, permitting the expansion of funding for stem cell lines derived after 2001. In the interim, a number of states stepped into the void, providing public funds for both derivation of new hESC lines and research. More difficult to trace is the private and foundation funding that was and still is available for stem cell research. The funding gave some states, California in particular, an edge in research that could be translated into clinical trials and marketable therapies. While the trials are still few and therapies are still merely on the horizon, there is an expectation that regenerative medicine using hESC will be available within the decade.

Additional readings

Original sources and other scholarly readings

Scholarly research requires reading original sources, including the speech that a president actually delivered, not a summary of the speech. Scholarly articles in peer-reviewed journals should be read in addition to the summaries provided in newspaper articles or on television news. Accessing all types of information is an important step in understanding an issue from many different perspectives.

Many scientific articles are available from the National Institutes of Health, National Library of Medicine (NLM). We have provided the NLM PubMed link for those articles that are available through the NLM. In most cases, the full text of the article is available. A tutorial is provided for those who might need assistance in accessing the material.

Presidential executive orders, speeches, radio addresses, and remarks are available to the public at a variety of government sites, including the Government Publishing Office (http://www.gpo.gov) and the National Archives (http://www.archives.gov). A number of universities maintain presidential papers. The University of California, Santa Barbara houses The American Presidency Project, which provides easy access to many presidential documents.

1. G. W. Bush, Address to the nation on stem cell research, August 9, 2001, in *Public Papers of the Presidents of the United States: George Bush, 2001*, Book 2, July 1–December 31, 2001, 954, Washington, DC: Government Printing Office, 2001; available at https://georgewbush-whitehouse.archives.gov/news/releases/2001/08/20010811-1.html.
2. G. W. Bush E.O. 13435 of June 22, 2007, Expanding approval of stem cell lines in ethically responsible ways. *Federal Register*, 72 (120): 34591–34593; https://www.gpo.gov/fdsys/pkg/FR-2007-06-22/pdf/07-3112.pdf.
3. G. W. Bush. Remarks on returning without approval to the Senate the Stem Cell Research Enhancement Act of 2007, in *Presidential Papers of the Presidents—George W. Bush*, Book 01, Presidential Documents—January 1–June 30, 2007, June 20, 2007, 775–777; https://georgewbush-whitehouse.archives.gov/news/releases/2006/07/20060719-5.html.
4. B. Obama, E.O. 13505, Removing barriers to responsible scientific research involving human stem cells. March 9, 2009. *Federal Register*, 74 (46), March 11, 2009: 10667–10668; https://www.gpo.gov/fdsys/pkg/FR-2009-03-11/pdf/E9-5441.pdf.
5. B. Obama, Remarks of the President—As Prepared for Delivery—Signing of Stem Cell Executive Order and Scientific Integrity Presidential Memorandum. https://www.whitehouse.gov/the-press-office/remarks-president-prepared-delivery-signing-stem-cell-executive-order-and-scientifi.
6. California, Text of Proposition 71, California Stem Cell Research and Cures Initiative; https://www.cdph.ca.gov/services/boards/HSCR/Documents/MO-Prop71-08-2007.pdf.
7. California Institute for Regenerative Medicine, About CIRM; https://www.cirm.ca.gov.
8. Text of Stem Cell Research Enhancement Act of 2005, and 2007, and 2009 and 2011. https://www.govtrack.us/congress/bills/109/hr810; https://www.govtrack.us/congress/bills/110/s5; https://www.govtrack.us/congress/bills/111/s487.

Secondary analysis and news articles

1. NIH, State Initiatives for Stem Cell Research; http://stemcells.nih.gov/research/state-research.htm.
2. D. E. Jensen, Evaluating California's stem cell experiment, *The Sacramento Bee*, November 15, 2014; http://www.sacbee.com/opinion/california-forum/article3924977.html.
3. J. P. Lefkowitz. Stem cells and the President—An inside account, *Commentary*, January 1, 2008, 125 (1):19–24, p. 21; https://www.commentarymagazine.com/author/jay-lefkowitz/.
4. California Institute for Regenerative Medicine; https://cirm.ca.gov.
5. N. Wade, Stem cell researchers feel the pull of the golden state, *New York Times,* May 22, 2005: B 5; http://www.nytimes.com/2005/05/22/us/stem-cell-researchers-feel-the-pull-of-the-golden-state.html.
6. J. W. Adelson and J. K Weinberg, The California stem cell initiative: Persuasion, politics, and public service, *American Journal of Public Health*, 100 (3), March 2010: 446–451; http://www.ncbi.nlm.nih.gov/pmc/articles/PMC2820047/.
7. A. B. Parson, The *Proteus Effect: Stem Cells and Their Promise for Medicine*, Washington, DC: Joseph Henry Press, 2004.
8. W. Wayt Gibbs, The California Gambit, *Scientific American,* June 27, 2005 at http://www.scientificamerican.com/article/the-california-gambit/.

Critical thinking activities

1. President Bush addressed the nation on stem cell research on August 9, 2001. Using his authority to educate the country about the benefits and costs of human embryonic stem cell research, he disturbed some and delighted others when he allowed federal research funding on a limited number of stem cell lines. Read the president's address at: http://www.archives.gov/federal-register/publications/presidential-papers.html. Next, read President Obama's remarks on March 9, 2009 removing all barriers to research. Write a five-page paper comparing each speech on three themes. Did you find any common themes in the speeches of both presidents? If so, what might explain these similarities? On what themes did the men diverge?
2. Congress, despite numerous attempts, was not able to pass legislation forcing President Bush to expand federal funding for stem cell research. Three versions of the Stem Cell Research Enhancement Act were introduced (H.R. 4682—109th Congress, H.R. 810—110th Congress, and H.R. 3—111th Congress). Go to http://www.congress.gov and find out what happened to these bills—were hearings held, was a vote taken, were amendments introduced. The bill that almost passed, H.R. 810, was vetoed by the President; the House failed to override the veto. Write a 5 page paper based on your thesis that overriding a presidential veto is difficult and why. Include in your discussion why it would have been easier to obtain a two-thirds majority in the Senate than in the House.
3. In this chapter we have focused on a number of states that decided to provide their own public funding for hESC research. California, a large and wealthy state, committed $3 billion to a 10 year effort. Visit the webpage of the California Institute for Regenerative Medicine (CIRM) at: http://www.cirm.ca.gov. Go to the Our Impact and drop down to Funding Clinical Trials. As you did when you researched clinical trials at NIH, pick three trials and write a three page paper discussing the conditions for participation, how many participants enrolled, money allocate and results. Some researchers have argued that if the federal government had fully funded embryonic stem cell research in 2001, California and other states would not have spent money duplicating work being done by NIH scientists. Write a two-page paper discussing the benefits or costs of scientific research duplication.

Notes

1. T. R. Dye, *American Federalism: Competition Among Governments,* Lexington, MA: Lexington Books, 1990.
2. K. J. Ryan, The politics and ethics of human embryo and stem cell research, in M. Ruse and C. A. Pynes (editors), *The Stem Cell Controversy,* 2nd ed., Amherst, New York: Prometheus Books, 2006: 291–300.
3. *Roe v. Wade* 410 U.S. 113, 1973. The Court held that a woman's right to an abortion fell within the right to privacy recognized in *Griswold v. Connecticut* 381 U.S. 479, 1965. The decision gave a woman total autonomy over pregnancy during the first trimester and defined different levels of state interest for the second and third trimester. As a result, the laws of 46 states were overturned by the Court's ruling.
4. Public Law 105-119 617, 1997 111 Stat. 2519, codified as a note following 18 U.S.C. SEC. 506. (a) None of the funds appropriated in this Act, and none of the funds in any trust fund to which funds are appropriated in this Act, shall be expended for any abortion. (b) None of the funds appropriated in this Act, and none of the funds in any trust fund to which funds are appropriated in this Act, shall be expended for health benefits coverage that includes coverage of abortion. SEC. 507 (a) The limitations established in the preceding section shall not apply to an abortion— (1) if the pregnancy is the result of an act of rape or incest; or (2) in the case where a woman suffers from a physical disorder, physical injury, or physical illness, including a life endangering physical condition caused by or arising from the pregnancy itself, that would, as certified by a physician, place the woman in danger of death unless an abortion is performed.
5. *Planned Parenthood of Southeastern Pa. v. Casey,* 505 U.S. 833, 846, 1992.
6. H. W. Jones, Jr., *In Vitro Fertilization Comes to America: Memoir of a Medical Breakthrough*, Williamsburg, VA: Jamestowne Bookworks, 2014.
7. M. J. Evans and M. H. Kaufman, Establishment in culture of pluripotential cells from mouse embryos, *Nature* 292, July 9, 1981: 154–156; also, G. R. Martin, Isolation of a pluripotent cell line from early mouse embryos cultured in medium conditioned by teratocarcinoma stem cells, *Proceedings of the National Academy of Sciences* 78, December 1981: 7643–7638.
8. A. B. Parson, The *Proteus Effect: Stem Cells and Their Promise for Medicine*, Washington, DC: Joseph Henry Press, 2004: 150–151.
9. M. Warnock, *A Question of Life: The Warnock Report on Human Fertilisation and Embryology*, Oxford: Basil Blackwell, 1985. Originally entitled *Report of the Committee of Inquiry into Human Fertilisation and Embryology*, July 1984.
10. Ibid., p. 84.
11. Merriam-Webster Dictionary at www.merriam-webster/com/dictionary/primative%20streak.
12. Canadian Royal Commission on New Reproductive Technologies, *Proceed with Care,* Ottawa, 1999. In 1989, a Royal Commission was set up to oversee licensing and monitoring of reproductive technologies. The Commission was chaired by Mary Baird and was often referred to as the Baird Commission.
13. Department of Health, Education and Welfare, Ethics Advisory Board. HEW support of research involving human in vitro fertilization and embryo transfer: Report and Conclusion, May 4, 1979.
14. Public Law 103-43, June 10, 1993, The National Institutes of Health Revitalization Act of 1993.
15. NIH, *Report of the Human Embryo Research Panel*, Vol. I at ix, 1994.
16. Ibid., xi.
17. W. J. Clinton, Statement. Federal Funding of Research on Human Embryos. *Public Papers of the Presidents of the United States*, Book 02, December 2, 1994: 2142.
18. *Planned Parenthood of Southeastern Pa. v. Casey* 505 U.S. 833, 846, 1992.
19. Public Law 104-99 110 Stat. 26, January 26, 1996, The Balanced Budget Downpayment Act. I.
20. W. J. Clinton, E.O. 12975, Protection of human research subjects and creation of national bioethics advisory commission, *Federal Register* 60 (193), October 3, 1995: 52063–52065.
21. NIH, National Bioethics Advisory Commission, Ethical Issues in Human Stem Cell Research—Volume 1—Report and Recommendations of the NBAC, August 1999; Volume 2: Commissioned Papers, January 2000; Volume 3; Religious Perspectives, June 2000.
22. Ibid., Vol. 1 Report and Recommendations of NBAC, August 1999, Executive Summary, p. 11.
23. H. Rabb. Letter from U.S. Department of Health and Human Services, Office of the General Counsel, Harriet Rabb to Harold Varmus, Director, National Institutes of Health, January 15, 1999; https://profiles.nlm.nih.gov/ps/retrieve/Narrative/MV/p-nid/191/p-docs/true.

24. NIH, National Institutes of Health guidelines for research using human pluripotent stem cells, *Federal Register*, 65 (166), August 25, 2000: 51975; http://stemcells.nih.gov/staticresources/news/newsArchives/fr25au00-136.htm.
25. R. Weiss, Nobel Laureates Back Stem CellResearch, *Washington Post*, February 22, 2001. http://www.washingtonpost.com/wp-dyn/content/article/2005/08/02/AR2005080201092.html.
26. P. John Paul II addresses President Bush, *American Catholic, News Feature,* July 23, 2001.
27. N. F. O'Brien, Embryonic stem-cell research immoral, unnecessary, bishops say, *Catholic News Service,* June 2008.
28. G. W. Bush, Address to the nation. Stem cell research. *Public Papers of the Presidents of the United States*, Book 02, July 1–December 31, 2001, August 9, 2001: 954.
29. Ibid.
30. NIH, Notice of Criteria for Federal funding of Research on Existing Human Embryonic Stem Cells and Establishment of NIH Human Embryonic Stem Cell Registry, November 7, 2001 NOT-OD-02-005, Office of the Director, NIH; at https://grants.nih.gov/grants/guide/notice-files/NOT-OD-02-005.html.
31. NIH, Stem cell and scientific progress and future research directions, June 2001. https//stemcells.nih.gov/info/2001report.htm.
32. NIH, Update on existing human embryonic stem cells, August 27, 2001; NIH, Federal Government Clearances for Receipt of International Shipment of Human Embryonic Stem Cells, November 16, 2001, NOT-OD-02-013; http://stemcells.nih.gov/policy/statements/pages/082701list.aspx.
33. NIH, Notice of criteria for federal funding of research on existing human embryonic stem cells and establishment of NIH human embryonic stem cell registry, November 7, 2001, NOT-OD-02-005, Office of the Director, NIH; https://grants.nih.gov/grants/guide/notice-files/NOT-OD-02-005.html.
34. Administrative Procedures Act (APA). Public Law 79-404, 60 Stat. 237, June 11, 1946. Federal regulations are created through a process known as rulemaking which is governed by the APA.
35. C. H. Cowan et al., Derivation of embryonic stem-cell lines from human blastocysts, *New England Journal of Medicine* 350, March, 2004: 1353–1356.
36. G. W. Bush, E.O. 13435, Expanding approval of stem cell lines in ethically responsible ways, *Federal Register* 72, June 20, 2007: 34591–34593.
37. NIH, Plan for Implementation of E.O. 13435, September 2007.
38. G. Q. Daley, Missed opportunities in embryonic stem-cell research, *New England Journal of Medicine* 351, August 2004: 627–628.
39. J. P. Lefkowitz, Stem cells and the President—An inside account, *Commentary* 125, January 1, 2008: 19–24, p. 21.
40. Ibid., p. 21.
41. Ibid., p. 20.
42. G. W. Bush. Message to the House of Representatives, returning without approval to the Senate the Stem Cell Research Enhancement Act of 2005. *Public Papers of the Presidents of the United States*, Book 02, July 19, 2006: 1425.
43. U.S. House of Representatives. H.R. 810 *Stem Cell Research Enhancement Act 2005*, 109th Cong, 1st Sess., and S.471; https://www.govtrack.us/congress/bills/109/hr810.
44. U.S. House of Representatives. H.R.3 *Stem Cell Research Enhancement Act 2007*, 110th Cong, 1st Sess.., and S.5; https://www.govtrack.us/congress/bills/110/s5.
45. C. Connolly, Frist breaks with Bush on stem cell research, *Washington Post*, July 30, 2005.
46. G. W. Bush, Message to the House of Representatives, returning without approval to the Senate the Stem Cell Research Enhancement Act of 2005, *Public Papers of the Presidents of the United States, George W. Bush*, Book 02, July 19, 2006: 1425.
47. Nightlight Christian Adoptions.
48. G. W. Bush, Message to the House of Representatives, returning without approval to the Senate the Stem Cell Research Enhancement Act of 2005. *Public Papers of the Presidents of the United States, George W. Bush*, Book 02, July 19, 2006: 1425.
49. G. W. Bush, Address to the nation. Stem cell research. *Public Papers of the Presidents of the United States*, Book 02, July 1–December 31, 2001, August 9, 2001: 954.
50. Pub. L. 109-242, 120 Stat. 570-571, July 20, 2006. Fetus Farming Prohibition Act of 2006.
51. Pub. L. 109-129 119 Stat. 2552, December 20, 2005. Stem Cell Therapeutic and Research Act of 2005.

52. G. W. Bush, E.O. 13435, Expanding approval of stem cell lines in ethically responsible ways, *Federal Register*, 72, June 20, 2007: 34591–34593.
53. G. W. Bush, Remarks on returning without approval to the Senate the Stem Cell Research Enhancement Act of 2007, *Public Papers of the Presidents of the United States*, Book 01, June 20, 2007: 775–777.
54. G. Meilaender, Stem cells and the Reagan legacy, *New Atlantis: A Journal of Technology and Society* 6, Summer 2004: 19–25.
55. Christopher and Dana Reeve Foundation.
56. The Michael J. Fox Foundation for Parkinson's Research.
57. Roman Reed Foundation: Hope for Spinal Cord Research.
58. Alliance for Regenerative Medicine.
59. B. Obama, E.O. 13505, Removing barriers to responsible scientific research involving human stem cells, *Federal Register* 74, March 11, 2009: 10667–10668; also, B. Obama, Remarks on signing an executive order removing barriers to responsible scientific research involving human stem cells and a memorandum on scientific integrity, March 9, 2009, *Public Papers of the Presidents of the United States*, Book 01, March 9, 2009: 199.
60. NIH, 2009 Guidelines on Human Stem Cell Research. In stem cell information, *Federal Register* 74, July 7, 2009: 32170–32175.
61. Ibid, 32172.
62. NIH, Human Embryonic Stem Cell Registry.
63. *Sherley v. Sebelius*, 610 F.3d 69, 73 (D.C. Cir. 2010); also, *Sherley et al. v. Sebelius et al.* Civ. No. 1:09-cv-1575 (RCL), August 23, 2010.
64. Pub. L. 104-99 110 Stat. 26, January 26, 1996, The Balanced Budget Downpayment Act, I.
65. *Sherley v. Sebelius*, U.S. Court of Appeals for the District of Columbia Circuit, No. 11-5241, 2012: 1–27, 11–12
66. M. Wadman, U.S. stem-cell chaos felt abroad, *Nature* 467, September 2010: 138–139.
67. Three bills were introduced in the House of Representatives, but only one had a Senate companion bill. *Stem Cell Research Advancement Act 2009, H.R. 4808. Cell Research Advancement Act of 2011. H.R. 2376*, Stem Cell Research Advancement Act of 2013, H.R. 2433, were all introduced by Congresswoman Diana DeGette (D-CO). While all three bills had bipartisan support, the majority of the cosponsors were Democrats. The Senate bill (S 3766) was introduced by Senator Arlen Specter (R-PA). During the 111th Congress (2009–2011), *the Stem Cell Research Enhancement Act 2009* was introduced again, but President Obama expanded the eligible stem cell lines for research. This Act was not needed, as its primary purpose was to extend the date of the eligible Bush lines (sometimes referred to as Presidential lines).
68. Alexander Hamilton, The same subject continued: The insufficiency of the present confederation to preserve the union, Federalist No. 17. *The Federalist Papers*.
69. Congressman Andy Harris (R-MD) introduced a bill in the 114th Congress. With only one sponsor, it has little chance of passing. *Human Cloning Prohibition Act of 2015, HR3498.*
70. *New State Ice Company v. Liebmann*, 285 U.S. 262, 1932.
71. According to the U.S. Department of Commerce, Bureau of Economic Analysis, the 10 states with the highest per capita income in 2003 were Connecticut, New Jersey, Massachusetts, Maryland, New York, New Hampshire, Minnesota, Colorado, California, and Illinois. Seven of these states would pass legislation to fund hESC research between 2004 and 2009. U.S. Department of Commerce, Bureau of Economic Analysis; http://www.bea.gov/.
72. W. J. Broad, U.S. Is Losing its dominance in the sciences, *New York Times*, May 3, 2004.
73. G. Jones, Bill boosting stem-cell research to be signed, *Los Angeles Times*, September 22, 2002.
74. California is a direct initiative state which means that proposals go directly on the ballot once a signed petition is presented to the secretary of state. Since Prop. 71 was an amendment to the State Constitution over 500,000 signatures were required to put the initiative on the ballot. Other states have indirect initiative meaning that the proposal is presented to the state legislature which can act on the measure and then, if needed (no action taken), put it on the ballot. National Conference of State Legislatures. http://www.ncls.org/research/elections-and-campaigns/chart-if-the-initiative-states.aspx.
75. A. Park, *The Stem Cell Hope: How Stem Cell Medicine can Change our Lives,* New York: Hudson Street Books, 2011.

76. J. Alberro, Economic Impact of Research Funded by the California Institute for Regenerative Medicine, March 2011.
77. H. B. Alberta et al., Assessing state stem cell programs in the United States: How has state funding affected publication trends? *Cell Stem Cell* 16, February 5, 2015: 115–118.
78. California Institute for Regenerative Medicine, *Clinical Trials.*
79. California Institute for Regenerative Medicine, *Beyond 2.0—Strategic Plan 2016 and Beyond.*
80. N. Wade, Stem cell researchers feel the pull of the Golden State, *New York Times*, May 22, 2005.
81. C. L. Jackson, State pulls back on stem cell funding, *New Jersey Com.*, June 22, 2008. http://www.nj.com/newark/index.ssf/state_pulls_back_on_stem_cell.html.
82. T. Somers, Defeat in N.J. of stem cell initiative raises alarm. *San Diego Union Tribune,* November 11, 2007. http://www.sandiegouniontribune.com/uniontrib/20071111/news_1b11njstems.html.
83. J. McDonald, Connecticut's continuing role in advocating stem cell research. *Yale Journal of Biology and Medicine* 82, September 2009: 97–99.
84. J, Palca, States take lead in funding stem-cell research, National Public Radio, March 30, 2007.
85. C. A. Cowan et al., Derivation of embryonic stem-cell lines from human blastocysts, *New England Journal of Medicine* 350, March 25, 2004: 1353–1356.
86. A. Park, *The Stem Cell Hope,* New York: Hudson Street Press, 2011: 133.
87. New York, Department of Health, New York State Stem Cell Science; New York Stem Cell Foundation.
88. Maryland Technology Development Corporation.
89. Illinois Regenerative Medicine Institute.
90. WiCell, http://www.wicell.org/home.
91. M. Baker, James Thomson: Shifts from embryonic stem cells to induced pluripotency, *Nature Reports Stem Cells,* August 2008; doi: 10.1038/stemcells.2008.118.

4 Of facts and frames

Sharing the news and influencing views

In 1998, when James Thomson isolated human embryonic stem cells, most people had no idea that it would not only revolutionize the field of cell biology, but also unleash a political and ethical firestorm. Most of us learned about the discovery from the media. Newspapers, magazines, and television became primary sources of information and speculation about the prospect that cures for degenerative diseases such as Parkinson's, Alzheimer's, and diabetes were just around the corner. Relying on unused embryos from IVF treatments, stem cell research also created a number of ethical and political controversies. While these controversies have not disappeared, the focus is shifting away from ethical concerns and toward the translation and safety of stem cell therapies.

In this chapter, we focus on the role of the media as a primary source of scientific information. The media serve as a conduit between scientists and the general public. The vast majority of the consuming public does not have the time or educational background to read scientific journals and tease out pertinent information. Projecting a timeline of how long it will take for a promising discovery to move from lab to clinic is even more daunting, so we rely on the media to inform us. In the process of informing and communicating, the media can also influence our views and attitudes on controversial issues. Stem cell research is one such issue. The second part of this chapter examines how polls and surveys, a regular feature in American life, have tried to assess public views about stem cell research. Most people knew little about the subject in 1998, but this changed once the issue moved to the top of the political agenda in 2001. Since then, national opinion polls began regularly asking participants about their views about stem cell research. Researchers and decision makers use these polls to gauge public support for legislation to either expand or limit research funding. Has the public changed its views about stem cell research? And if so, what is the reason for the change?

How do most of us find out about scientific and/or medical issues? Unless you have a scientific background or are especially curious about a medical procedure, you learned about the issues from reading newspapers, watching TV, or surfing the Internet. For most of us, the media are the main sources of scientific and technological information. How the media present or frame an issue can influence how we perceive that issue. While it is generally true that the media present us with the facts in the case, it is also true that facts can be framed in a way that emphasizes or favors one particular view.[1] In short, the words, phrases, and stories used by the media can shape our perceptions. We might be predisposed to support or oppose a certain issue, but the media can help push us firmly into one camp or the other. As we noted in earlier chapters, some people are of the opinion that stem cell research was, and still is, linked with abortion. If a person is opposed to abortion, then a news article about hESC research that mentions the destruction of embryos might be a cue for the reader to establish opposition to the measure.

The use of frames and the concept of framing are used across many disciplines, including the sciences. Scientists are even encouraged to "bridge the divide between science and journalism" by, among other things, communicating simply and clearly with reporters.[2] In 1993, Robert Entman set out his theory of media frames by noting that frames "select some aspects of a perceived reality and make them more salient in communicating text, in such a way as to promote a particular problem definition, causal interpretation, moral evaluation, and/or a treatment recommendation."[3] The frame helps the reader relate the topic to his/her own view of reality. In this way the frame would function as a dependent variable.[4] The frame can also be used by the journalist, scientist, or politician to project a particular reality to the audience. A frame organizes the message for the reader and, by highlighting specific facts or values, frames influence how the reader perceives an issue. In this case, the frame would function as an independent variable.[5]

This use of frames becomes even more important when studying how scientific issues are covered, because "the media are the sole providers of information about science and technology for a very large segment of the population."[6] Most of us do not have access to technical journals, scientific researchers, and current data to help shape our perceptions of scientific issues. And even if we had access, would we take the time to become thoroughly informed? According to research conducted by Nisbet, the answer is no.[7] He concludes "that the public by nature is 'miserly', with individuals relying on their value predispositions and only the information most readily available to them from the mass media and other sources in order to formulate an opinion about science controversy."[8] Nisbet's theory is supported, in part, by a poll conducted by the Pew Research Center that asked the public which source had the biggest influence on their opinions about government funding of stem cell research. In 2002, 34% said that the media were the biggest influence, compared with 22% who answered education and 18% who said religious beliefs. However, when Pew compared the responses based on whether the person thought that government should fund stem cell research, the percentage that cited the media as being the biggest influence increased to 41%. In contrast, media influence dropped to 27% for those respondents who said that government should not fund stem cell research, while the importance of religious beliefs increased to 34%.[9] In another survey conducted by Pew in July 2005, the findings were similar. If respondents believed that stem cell research would result in new medical cures, they reported that the media and education had the biggest influence. If respondents believed that protecting the potential for life was more important than conducting stem cell research, they reported that religious beliefs had the biggest influence.[10] It is a little easier to speculate that religious beliefs predisposed the individual to oppose stem cell research. It is a little harder to suggest that reading about future stem cell cures leads an individual to support the research. It may be that people who are already predisposed to support medical breakthroughs search for these stories in the media.

Patrick Hopkins performed another study that examined the role of the media in influencing opinion.[11] He focused on how the media represent cloning (an issue closely linked to stem cell research), and noted that "[m]edia coverage fixed the content and outline of the public moral debate, both revealing and creating the dominant public worries about cloning humans."[12] Hopkins concluded that "the media … influence what we think and talk about, what we take to be important, what we worry about."[13] He looked at the coverage of cloning as portrayed in the popular media (*Time Magazine*, *Newsweek*, *U.S. News & World Report*), after the birth of Dolly, the cloned sheep. While the "potential medical and agricultural benefits are usually mentioned," Hopkins also notes, "These benefits, however, are always juxtaposed to the dangers of cloning in alarmist, emotion-packed ways."[14] He makes another point worth sharing that illustrates how a phrase or, in this case, a book title, can trigger a values cue for the reader. "Most people have never read *Brave New World*, but that doesn't matter. The scores of references to *Brave New World* aren't about the book; they are about the trope connected to the book. *Brave New World* is a stand alone reference, image, and warning … epitomized and made possible by the technology of cloning."[15] Even if you have not read *Brave New World*, think about the image that the title conveys. The image that comes to mind is an example of a frame. In his August 2001 speech on stem cell funding, President George W. Bush also used the *Brave New World* frame when he acknowledged that "We have arrived at that brave new world that seemed so distant in 1932 when Aldous Huxley wrote about human beings created in test tubes in what he called a hatchery."[16] What image comes to your mind when you read the phrases "test-tube babies" or "hatchery"? These and other phrases found in articles about hESC research might support or challenge one's views about the research or, assuming the person has read nothing about the research, contribute to the formation of an opinion.

A number of studies have applied the concept of news frames to analyze how newspapers in the United States and other countries present the stem cell research debate.[17] For the most part, we will discuss studies that focus on the U.S. media. Studies that use different methodologies will introduce the reader to the different ways that researchers try to determine how and whether news articles (frames) influence public opinion.

Until August 2001, when President Bush gave a speech to the nation on stem cells, most people did not know much about hESC or the prospects for medical cures. Public opinion polls show that the general public was not informed.[18] So it is logical that studies looking at media frames before August 2001 found that frames had little impact on the readers.

Before we move to these studies, it is helpful to look at the number of articles printed in four of the major U.S.-based newspapers between 1998 and 2016 (the *New York Times*, the *Washington Post*, *USA Today*, and the *Wall Street Journal*). In Table 4.1, we report the results of our content analysis of articles using the search term "stem cell research." Our quantitative findings are informed by some of the key events in U.S. political life, which help explain why there are more articles in certain years and fewer in others.[19]

One of the first studies to apply the framing heuristic to stem cell research was published in 2003. The authors analyzed stem cell–related articles appearing between 1975 and 2001

Table 4.1 Media coverage of stem cell research

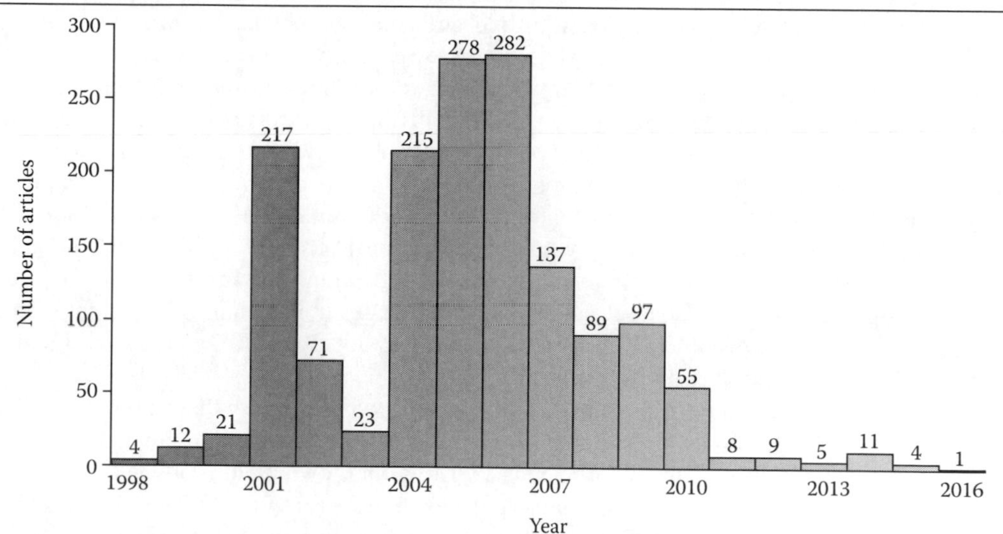

Date	*Event*
2001	President Bush speech—limiting funding
2004	Bush/Kerry election; California Prop. 71; President Reagan dies; South Korean stem cell (SC) fraud; Ron Reagan speech at Democratic Convention
2005	Congress expands SC funding; Bush threatens veto
2006	Bush vetoes SC funding bill
2007	Bush issues E.O.—limited SC funding
2008	President Obama elected
2009	Obama E.O.—expands funding
2010	Midterm elections: candidates differ on SC
2014	Japanese SC research fraud

Source: Adapted from the *New York Times*, *Washington Post*, *USA Today*, and *Wall Street Journal*.

in the *New York Times* and *Washington Post*.[20] Although stem cell research received only modest media attention before 2001, it peaked in 2001 after President Bush's speech. While acknowledging that more research was needed, the study concluded with three findings that contributed to the literature of framing and science. First, they noted that the policy context matters. There was limited media attention on the stem cell issue until President Bush and Congress collided. Initially, the controversy was contained within the NIH and HHS administrative arenas, where technical issues could be resolved among experts. In short, there was nothing to dramatize before 2001. Second, attention increases when journalists can cover new events, "using recycled thematic formats and storytelling conventions."[21] In 2001, new stories about stem cell research could be linked to stories about "previous controversies surrounding abortion and fetal transplantation."[22] Finally, media attention increases when there is a potential for the issue to be "framed in dramatic terms."[23] Media interest peaks when a story can be presented as political or ethical conflict; the public is more likely to pay attention to a controversial issue. If the public pays attention, then it is reasonable to assume that an opinion will be formed.

In this finding, the implication is that the public is already predisposed to view a new issue (stem cell research) based on the link to an old issue (abortion). The media can reuse the language and frames from their reporting on abortion controversies and apply them to the stem cell controversy. What is on the front page will change, depending on the players. When one of the players is the president, there is a high probability that the story will receive maximum media attention and become front-page news rather than be buried in the back pages.

In a related study published in *Science*, the authors discuss the intersection of science and politics and challenge scientists to actively "frame" information to make it relevant to different audiences. They note that "frames organize central ideas, defining a controversy to resonate with core values and assumptions."[24] Many scientists retain the well-intentioned belief that, if laypeople better understood technical complexities from news coverage, their viewpoints would be more like scientists', and controversy would subside. On the embryonic stem cell issue, patient advocates have delivered a focused message to the public, using social progress and economic competitiveness frames to argue that the research offers hope for millions of Americans. These messages helped drive up public support for funding between 2001 and 2005. However, opponents continue to frame the debate around the moral implications of research, arguing that scientists are playing God and destroying human life. Ideology and religion can screen out even dominant positive narratives about science, and reaching some segments of the public will remain a challenge.

The same authors published an editorial in the *Washington Post* with the provocative title, "Thanks for the facts: Now sell them," in which they implore scientists to create their own frames.[25] Is there anything wrong with scientists writing their own media copy?

In another study examining media frames and stem cell research in the United States and Brazil between 2001 and 2005, Reis found thematic differences between the two countries.[26] In the United States, 86% of the news articles focused on political or ethical/religious issues, compared with only 44% of the stories in Brazil, while 60% of the articles in the Brazilian press focused on medical/scientific frames, as opposed to only 35% of those in the United States.[27] Reis engaged in the same kind of contextual analysis that is typical of research using this method. He selected the papers, gathered the articles, read them, and assigned them to categories based on the dominant themes he and his coders identified. It is typical that more than one individual read the news articles and assigned them to a category or frame. This design promotes research reliability. Researchers are looking for key words or phrases that clearly categorize an article. An easy way to categorize frames might be to classify them as positive, neutral, or negative. In this study, Reis identified four themes: medical/scientific, political, ethical/religious, and business. He noted that in U.S. newspapers even those articles that were classified as scientific/medical used words like "battleground," "deeply mired in controversy," or "extremely polemical."[28] Articles reporting on scientific coverage typically used case studies to give readers a more illustrative or relatable account of the subject. The following excerpt, taken from a 2004 *Washington Post* article, is an example of how news frames can personalize a story:

> Nancy Reagan, furthermore, has championed the cause of Alzheimer's patients with the kind of clout that few other caregivers could wield, and the Reagan name has helped raise millions for research. Nancy Reagan has also led the fight against federal restrictions on embryonic stem cell research—discreetly challenging President Reagan's most prominent admirer, President Bush, who imposed the restrictive policy.[29]

Most people reading this article recognize Nancy Reagan, an admired first lady and an outspoken advocate for stem cell research. She and President Reagan, who suffered from Alzheimer's later in life, engaged in a public debate about the struggles faced by patients and caregivers. Personal stories can familiarize audiences with different aspects of the debate. News coverage in the United States is also driven by national events. The 2001–2005 period chosen by Reis to analyze media frames coincided with very significant political developments in the United States. In 2001, President Bush issued his position in favor of limiting funding for hESC research. The 2004 presidential election juxtaposed President Bush's position limiting hESC research with Democratic candidate John Kerry's support for expanding funding. Voters in California approved a $3 billion initiative (Prop. 71 in November 2004) to fund the state's own hESC research efforts. President Reagan passed away in June 2004. His son, Ron Reagan, gave a speech at the Democratic National Convention asking the nation to "cast a vote for embryonic stem cell research."[30] Each of these events, discussed in news articles, could contribute to an individual's assessment of the pros and cons of stem cell research.

In 2005, Congress began debating the Stem Cell Research Enhancement Act of 2005, which was eventually vetoed by President Bush. But in the interim, news articles reported on the showdown between the president and Congress. By now you probably see the connection between political events and media attention. An interesting research question is to what extent the media frame contributes to an individual's attitudes toward stem cell research, or does an individual's attitude (developed by a variety of factors) explain what he or she extracts from a news story? In order to answer these questions, researchers need to survey individuals. Opinion polls, as well as more elaborate research models, are the usual vehicles for collecting information about individuals and how they react to specific media frames.

Before we move on to polls, it is worth discussing a recent study that examined media frames before and after the biopharmaceutical company Geron terminated its clinical study using hESC to repair spinal cord injuries in November 2011. In this study the event is more scientific than political.[31] Researchers focused on how the media portrays the issue from January 2010 to December 2013. Geron is the U.S. company that financed the research of both Thomson and Gearhart in exchange for patent rights to hESC. The company won U.S. Food and Drug Administration (FDA) approval in 2009 to begin clinical trials on spinal cord injuries using hESC therapies. News articles reporting the story implied that if mice could walk again after injections of hESC, then it might be possible for humans to do the same. After just 1 year and 5 patients, the company terminated the study due to financial considerations. Because Geron conducted the first clinical trial using hESC, it was widely reported. You can see the spike in news articles in Table 4.1.

After analyzing 307 articles (the study included publications in Canada and the United Kingdom), the authors found that the overall perspective on the future of stem cell research was optimistic in 58% of the publications, neutral in 32%, and pessimistic in 10%. In addition to the media's optimistic slant (frame), Kamenova and Caulfield found that the predicted timeline for therapies pre- and post-Geron did not change substantially. "Whereas only 10% of the pre-Geron predictions indicated an expectation of SC therapies in the distant future, this number increased to 19% in the post-Geron data."[32] The termination of the Geron trial, in addition to the fact that there were no other hESC trials in 2013, may have contributed to the increase in the duration of the projected timeline.

The authors discovered that the major shift was "from ethical, legal, and social issues, which were central to media framing and public debates in the past, to stories about clinical translation and new discoveries."[33] This should not be surprising, given that President Barack Obama's order to expand hESC research reduced the political tension between the president and Congress. Table 4.1 confirms a big drop in news articles after President Obama signed his executive order expanding research funding.

Another way to study media frames and their effect on individuals is to use an experimental design. This is a common method to study the effectiveness of experimental drugs, but it is also used in the social sciences. You need at least two groups of participants; one group undergoes the experimental treatment, while the other receives a placebo (nothing). Then the results are compared. Since it is almost impossible to control all aspects of a participant's environment, researchers create groups based on some common factor. For example, we might compare students taking a college course given in a lecture format with students taking the same course using an online format and then determine which group learned more. In the social sciences, this is referred to as a quasi-experimental design.

By employing this type of methodology, researchers can give study participants news articles, ask them to read the articles, and then have them answer (written or oral) questions about the articles. In this way, the researchers can discern whether the frame used in the articles influenced the participants' views on an issue.

Researchers at a U.S. university used such an experimental design to investigate what effect selected news frames had on an individual's ratings of research using hESC compared with adult stem cells.[34] Researchers wanted to see what effect a slanted (framed) news article would have on the participants' views of the research. Sixty college students were divided into three groups. Each group was given an article from the *New York Times* rewritten by the researchers to emphasize three different frames—a political conflict frame, an economic prospect frame, and a scientific prospect frame. As in any good research design, the investigators also asked students many demographic questions (including religion, major, and political orientation). After reading the assigned article, the students were asked questions designed to detect whether and to what extent the article's frame influenced their assessment of hESC versus adult stem cell research.

The study found that, in terms of the political conflict frame (ethics), participants made a significant distinction between hESC and adult stem cell studies. Participants rated adult stem cell research as more ethical than hESC, however, there was no difference in how participants rated the different stem cell types in terms of credibility (economic prospects), and usefulness (scientific prospects). Unsurprisingly, the study found that the independent variable religion had the greatest effect on whether respondents assigned a positive value to the political conflict frame.[35]

In another article using the same data as that discussed above, the researcher found little difference in opinion between science majors and nonscience majors on the ethicality (political frame) of research using hESC versus adult stem cells. However, science majors reported that research with hESC was more credible (economic frame) and more useful (scientific frame) than did nonscience majors.[36] The findings suggest that university major acts as an independent variable that influences how an individual perceives information conveyed in a news article.

Panel data is another design used by researchers to determine how individuals' views change over time. Panel surveys are expensive to administer, but the opportunity to query the same people multiple times contributes to a better understanding of how and why change occurs. Researchers used a study that included panel data from a Lifestyle Study conducted for the advertising agency DDB to investigate whether public attitudes about hESC are influenced by individuals' value predispositions or cues from the media.[37] The study used data from a three-wave panel survey administered between 2001 and 2005. A single panel (group) of respondents was interviewed in 2002, in 2004, and again in 2005. One of the difficulties with panel studies is attrition—people disappear. The study started with over 4000 respondents in 2002; by 2005 slightly over 1000 remained in the study.[38] However, this is still a sizable sample.

The findings reveal that attitudes toward stem cell research were formed by value predispositions rather than cues from the news media. The other variables that affected an individual's predisposition were religiosity, ideology, and deference to scientific authority. Scientific knowledge was found to play a limited role in individual attitudes toward stem cell research. Among highly religious individuals, on the other hand, opposition to stem cell research was strong.[39] One benefit of a large survey is the opportunity to test a number of hypotheses using sociodemographic variables that are routinely collected in these studies. What we learn from this study is that religion plays an important role in determining individual attitudes about stem cell research. This may not come as a big surprise; studies that focus on individual beliefs about abortion also found religion to be a key explanatory factor.[40]

Polls and surveys: Taking the public's policy pulse

In the first part of this chapter, we looked at theories about whether the media inform and influences the public about current policy issues. Examining the different ways in which the media frame a policy issue helps us gauge the issue's importance. A larger number of articles written about an issue suggests issue salience in public discourse. The more controversial an issue is (disagreement between the president and Congress—the House and Senate, organized interest groups), the more newsprint will be allocated to that issue. The more articles written about an issue, the more likely the public will read the articles and form an opinion. At least, that is the theory. In Table 4.1 we linked media attention to key political and scientific events. But just because the media think an issue is important does not necessarily mean the public feels the same way.

Public opinion polls have become another way for researchers to gauge public sentiment about an issue. Polls are also a way for elected officials to find out what is on the public's mind. What does the public support and what does it oppose? Should I vote to pass a bill or should I vote against the proposal? This decision is much easier when polls indicate that 80–90% of the population favor an issue. It is not so easy when the figures indicate that the population is evenly split or when those in favor account for 60% of the population. Politicians are avid consumers of opinion polls. It is common for politicians and other decision makers to commission their own polls, however, commissioned polls can be problematic, especially if the questions are skewed to elicit a preferred response, or if the sample size is too small and not representative. Developing, administering, and analyzing poll data requires preparation and experience.

There are many kinds of polling. Marketing research is one type of specialized polling used to figure out what individuals like about a product and how best to advertise the product for maximum sales. These polls target respondents based on demographics such as age, gender, or income. Alternatively, the sample might query only those people who already use the product. Marketing polls are paid for by companies engaged in commerce and the purpose of the survey is to find out how to sell more of their goods or services.

Election polls are another common type of poll that typically emerge in full force before and after elections. These polls (national or statewide in scope) are interested in who the respondents plan to vote for in the upcoming election and why. Election polls target individuals over 18 (voting age) and/or those who are registered to vote. The latter tends to be a better predictor of actual voting. These polls are paid for by candidates or their campaign committee, or by news organizations eager to predict the winners ahead of the competition.

Public opinion surveys are an outgrowth of both marketing and election polls. These polls are designed to monitor public views on a variety of issues and attitudes. The polls also inform the public about developing trends relative to salient issues. These polls are used by policy makers, civic leaders, and the public to assess changes in thinking and analyze how those changes affect policy decisions. Who pays for these polls? They are often part of ongoing data collection by the large polling companies, such as Gallup. These organizations (discussed below) publish their findings in research reports and sell their data to government agencies, academics, and news organizations. Polling companies also conduct paid surveys for groups interested in learning more about their members or employees. If a group is interested in an occasional survey, then hiring an established polling company is more cost-effective. The field has become so specialized that even large polling organizations will contract out parts of the survey process. Typically, sample selection and phone interviews are purchased on an as-needed basis.

Polling was once the monopoly of a few large organizations such as Gallup, Harris, and Roper; today, however, there are hundreds of companies, large and small, that are paid to measure the public's pulse. A few words about polling are in order. Most of us have participated in a poll or survey. Some are short and simple, asking the respondent to rate the service of a restaurant or the bank. Others are more elaborate and often include demographic questions about your age, gender, religion, income, political party preference, education, and so forth. Polls gather information on: what people believe, how they feel, and how they will act. From that information, the researcher can extrapolate an outcome, plot a course of action, or just publish the findings. Policy makers can then make their own assessments. Depending on the results of the survey, policy makers can use the data to support their position; if they do not like the results, they can ignore the data or find another survey that does support their preferred policy position. We will discuss the existence of contradictory survey results later on.

Among the key factors in evaluating the reliability of a poll is the sample size, who was interviewed, how and when the data was collected, and, finally, the margin of error. A television newscaster reporting on election polls will usually tell you that the margin of error is, for example, plus or minus 3%. Then you can figure out how accurate the poll is by adding or subtracting 3% from the numbers for Candidate A and Candidate B. Pollsters will publish (online or in written reports) the size of the sample and the exact wording of the questions.

Most methodology books include a chapter or two on survey research.[41] Polling has become a profession, and an increasingly technical one. Here we will rely on information from Gallup to give you an overview of some of the conventions used in poll construction.[42] In 2001, Gallup introduced the Gallup Poll Social Series (GPSS), which are still in use today. In this series adults are interviewed annually on a number of social, economic, and political issues. In 2001, Gallup included a question on embryonic stem cell research, and they continue to ask the same question today using the same wording. The size of this sample is typically about 1000 respondents selected at random. The key word here is "random," to ensure that the poll captures the opinions of a representative sample of the U.S. population (or whatever population the polls want to assess). If we just want to know what college students think about hESC research, then we want a randomly selected sample of college students from across the country. Gallup interviews adults over 18 years of age living in all 50 states and Washington, DC. Most interviews are conducted by phone. However, beginning around 2005, Gallup, like other polling organizations, began to use a dual-frame design that includes both landline and cell phone users in the sample. As more people became cell phone-only users, pollsters had to adjust the research design to make sure they had a random sample. If you are thinking that pollsters should start doing surveys online, then you will not be surprised to find that some already do. What are some possible issues with internet-only surveys?

Two other factors that might influence the outcome of a poll are timing and wording: When was the survey administered, and how is the question worded? In order to compare data over the course of many years, the question must be exactly the same. A survey conducted *before* President Bush's 2001 televised address to the nation may yield very different results than a survey conducted after the speech. A person who watched the speech might have a better understanding of what stem cells are, their potential, and the ethical issues surrounding the research. If the survey was conducted several years later, results might revert back to the pre-speech findings, as the issue might have faded from the media as well as the public's attention.

Gallup is one of the oldest and most well-known polling organizations. It was also among the first to include questions about stem cell research in its polls. Other groups that pose stem cell questions include the Virginia Commonwealth University (VCU) Life Sciences Survey, the Pew Research Center, and the Kaiser Foundation. Other organizations included questions about the subject in only one or two surveys, changed the wording, or used the Internet to query respondents.

We begin with Gallup which, in 2001, initiated the Gallup Poll Social Series, a set of public opinion questions designed to monitor U.S. adult views on a number of key issues. A question on stem cell research was introduced in 2001 and has been asked annually ever since. The question underscores the morality of using embryonic stem cells in research and emphasizes its focus on the respondent's personal beliefs. There is no right or wrong answer, and the question does not add any additional information or mention other types of stem cells used in research.

Table 4.2 indicates that a solid and consistent majority of respondents feel that the research is morally acceptable. The percentages have increased over the 13-year period but it is not an overwhelming majority. Would a policy maker looking at this information vote to expand funding for research using human embryonic stem cells?

Given that most articles that discuss hESC also mention potential cures for Alzheimer's, Parkinson's, cancer, and heart disease, Gallup analyzed the data from their surveys by age groups: 18–34, 35–54, and 55 and older. The findings in Table 4.3 indicate that older respondents initially viewed stem cell research as less morally acceptable than did their younger counterparts. However, as Gallup notes, "[a]cceptance among this older age group has risen by 20 points over the past 11 years."[43] Acceptance of stem cell research has leveled off among all age groups. Why do you think older adults are now more accepting of stem cell research than they were in 2001?

Table 4.2 Embryonic stem cell research morally acceptable 2001–2005
Q. Next, I am going to read you a list of issues. Regardless of whether or not you think it should be legal, for each one, please tell me whether you personally believe that in general it is morally acceptable or morally wrong. How about medical research using stem cells obtained from human embryos?

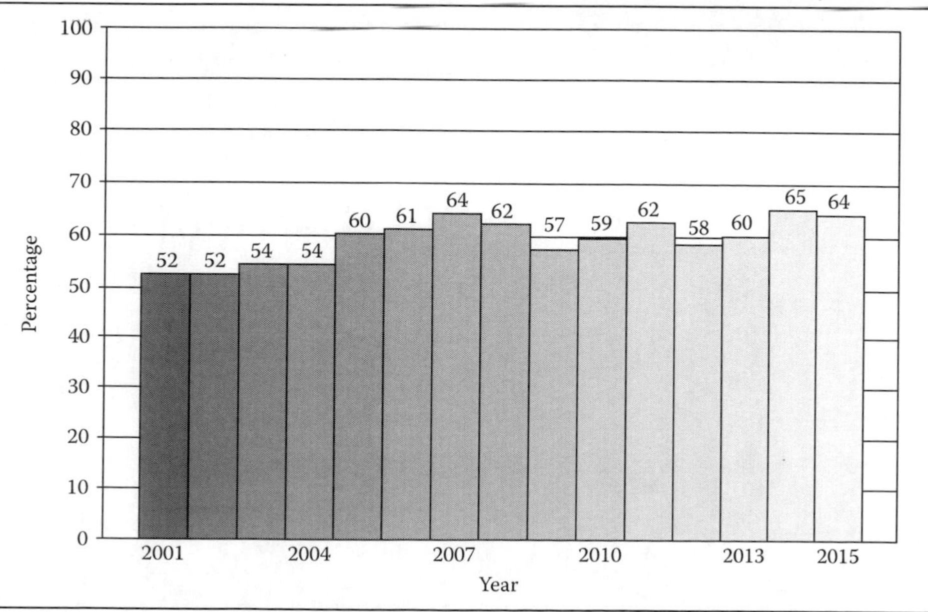

Source: Adapted from Gallup Poll Social Series.

Table 4.3 Embryonic stem cell research morally acceptable by age

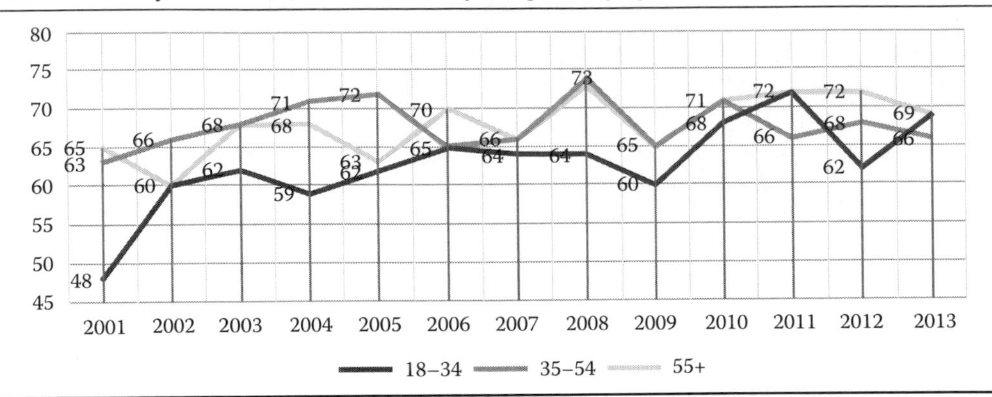

Source: Adapted from Gallup Poll Social Series.

Between 2001 and 2010, Virginia Commonwealth University (VCU) administered their annual Life Sciences Survey that included a number of questions about stem cell research (they no longer conduct the survey). Table 4.4 reports the findings from the question on the VCU survey that comes closest to the wording used in the Gallup data. The VCU survey asked respondents: "On the whole, how much do you favor or oppose medical research that uses stem cells from human embryos? Do you strongly favor, somewhat favor, somewhat oppose, or strongly oppose this?" We collapsed the two categories into "favor" and "oppose." This is typically done with questionnaires that use Likert-scale response categories.[44] The rate of support (favor) for embryonic stem cell research recorded in the VCU survey was much lower. In fact, in 2002, only 35% (down from 48% in 2001) of the VCU respondents said they favored the research,

Table 4.4 Medical research using embryonic stem cells
Q. *On the whole, how much do you favor or oppose medical research that uses embryonic stem cells? Do you strongly favor, somewhat favor, somewhat oppose, or strongly oppose?* **Percent who strongly or somewhat favor.**

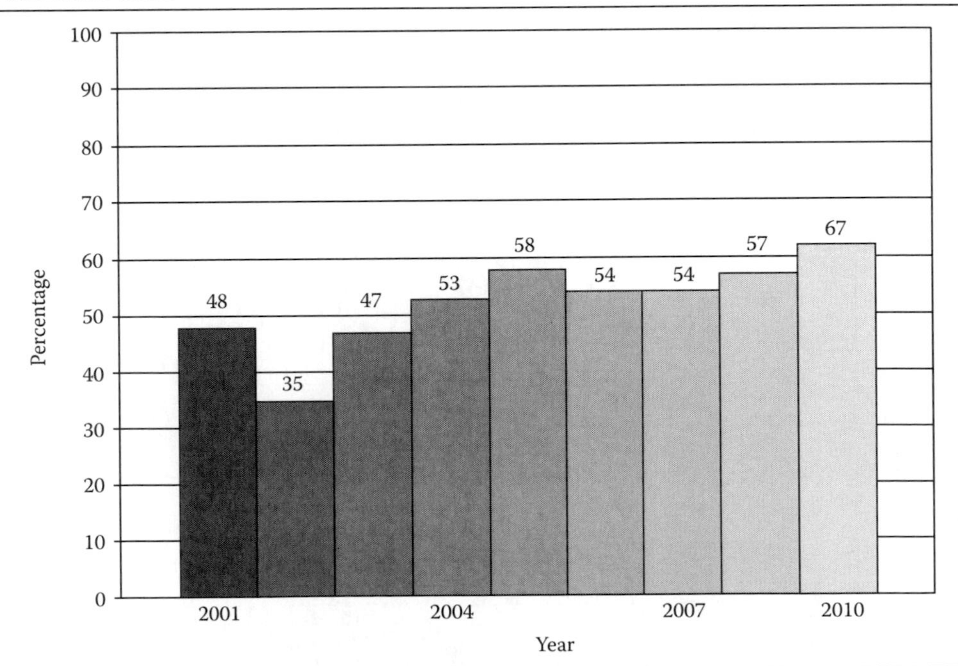

Source: Adapted from Virginia Commonwealth University Life Science Surveys.

Table 4.5 Support for stem cell research 2002–2013
Q1. *All in all, which is more important conducting stem cell research that might result in new medical cures, or not destroying the potential life of human embryos involved in research?*
Q2. *Do you personally believe that embryonic stem cell research is morally acceptable, morally wrong, or is not a moral issue?*

	2013 Different question asked				
	2002	*2004*	*2006*	*2009*	*2013*
Conduct	43	56	53	54	68
Not Destroy	38	32	32	34	22
DK	19	12	15	14	9

Source: Adapted from Pew Research Center.

compared to 52% of the respondents from the Gallup poll. What do you think might be some reasons for this difference? (Hint: in 2001, the VCU poll was conducted in September.)

In their own effort to gauge public opinion, the Pew Research Center also included questions on the stem cell issue from 2001 to 2009. The questions were part of a larger survey on U.S. politics. In 2013, Pew included a newly worded question on stem cell research as part of a survey on religion and politics. The question most similar to the Gallup and VCU polls asked respondents: "All in all, which is more important: conducting stem cell research that might result in new medical cures or not destroying the potential life of human embryos involved in this research?" Table 4.5 indicates that in 2002 a higher number of respondents said that not destroying an embryo was more important, 38%, compared with 32% in 2004. The number of respondents who believed that conducting research is more important increased from 43% in 2002 to 56% in 2004 and 54% in 2009. The greatest increase occurred in 2013, when 68% of the respondents indicated that it was either morally acceptable or not a moral issue to conduct stem cell research. However, it is important to note that the wording of the question changed in 2013 (Table 4.5). Is it acceptable to compare the findings of the 2013 survey with that of earlier surveys given the change in the wording of the question? Presented with both questions, would your response be different or the same?

So far, none of the polls discussed asked respondents about federal funding for stem cell research. As you may recall, President Bush's 2001 address concluded with his decision to fund federal research on a limited number of stem cell lines. There never was, nor is there today, a limit on how much the private sector or state governments can spend on hESC research. How did the public feel about taxpayer funding? In joint studies conducted by CNN, *USA Today*, and Gallup, that question was asked as early as 2005. The findings presented in Table 4.6 indicate steady support that is over 50% between 2005 and 2010. But that means that almost as many respondents felt that the federal government should not fund the research or had no opinion. Results like these are unsatisfying if you are a policy maker trying to decide how to vote.

In July 2005, President Bush vetoed the Stem Cell Research Enhancement Act of 2005 (H.R. 810/S471), passed by both houses of Congress, despite considerable public support based on the polls that we have just reviewed. The support was, however, not overwhelming in favor of research. Gallup provides insight into how strongly the public feels about an issue. In May 2005 respondents

Table 4.6 Federal funding of embryonic stem cell research
Q. *Do you think the federal government should or should not fund research that would use newly created stem cells obtained from human embryos?*

	Aug. 2005	*Aug. 2006*	*May 2007*	*Sept. 2010*
Should fund	56%	51%	53%	55%
Not fund	40%	41%	41%	41%
No Opinion	4%	8%	6%	4%

Source: Adapted from CNN/*USA Today*/Gallup Trend.

7. M. C. Nisbet, The competition for worldviews: Values, information, and public support for stem cell research.
8. Ibid, p. 90.
9. The Pew Research Center, *Cloning Opposed, Stem Cell Research Narrowly Supported: Public Makes a Distinction on Genetic Research* April 9, 2002, http://www.people-press.org/files/legacy-pdf/152.pdf.
10. The Pew Research Center, *Strong Support for Stem Cell Research: Abortion and Rights of Terror Suspects Top Court Issues*, August 3, 2005, http://www.people-press.org/files/legacy-pdf/253.pdf.
11. P. D. Hopkins, Bad copies: How popular media represent cloning as an ethical problem, *Hastings Center Report* 28 March–April 1998: 6–13.
12. Ibid., p. 6.
13. Ibid., p. 6.
14. Ibid., p. 11.
15. Ibid., p. 11.
16. G. W. Bush, Address to the nation. Stem cell research, *Public Papers of the Presidents of the United States*, Book 02, July 1–December 31, 2001, August 9, 2001: 954.
17. M. C. Nisbet et al., Framing science: The stem cell controversy in an age of press/politics, *The International Journal of Press/Politics* 8, April 1, 2003: 36–70; R. Reis, How Brazilian and North American newspapers frame the stem cell research debate, *Science Communications* 29, March 2008: 316–334; K. Kamenova and T. Caulfield, Stem cell hype: Media portrayal of therapy translation, *Science Translational Medicine* 7, March 11, 2015: 1–4; D. Nelkin, *Selling Science: How the Press Covers Science and Technology*, New York: W.H. Freeman & Company, 1995.
18. M. C. Nisbet, Public opinion about stem cell research and human cloning, *Public Opinion Quarterly* 68, Spring, 2004: 131–154; M. C. Nisbet, The polls—Trends: Public opinion about stem cell research and human cloning, *Public Opinion Quarterly* 68, Spring 2004: 132–155; M. C. Nisbet and A. B. Becker, The polls—Trends: Public opinion about stem cell research, *Public Opinion Quarterly* 78, Winter 2014: 1003–1022; M. Nisbet, Understanding what the American public really thinks about stem cell and cloning research, *Science and the Media*, The Committee for Skeptical Inquiry, May 2004; M. Nisbet and E. M. Markowitz, Understanding public opinion in debates over biomedical research: Looking beyond political partisanship to focus on beliefs about science and society, *PLoS One* 9, 2014.
19. We used LexisNexis to compile all of the articles published between 1998 and 2016 using the search term "stem cell research." We excluded letters to the editor but retained editorials and op-ed articles from syndicated columnists or guest articles.
20. M. C. Nisbetet al., Framing science: The stem cell controversy in an age of press/politics, *The International Journal of Press/Politics* 8, April 1, 2003: 36–70.
21. Ibid., p. 65.
22. Ibid., p. 66.
23. Ibid., p. 66.
24. M. C. Nisbet and C. Mooney, Framing Science.
25. M. C. Nisbet and C. Mooney, Thanks for the facts: Now sell them, *Washington Post* April 15, 2007.
26. R. Reis, How Brazilian and North American newspapers frame the stem cell research debate, *Science Communications* 29, March 2008: 316–334.
27. Ibid., p. 326.
28. Ibid., p. 331.
29. S. Vedantam, Reagan's experience alters outlook for Alzheimer's patients, *Washington Post* June 14, 2004.
30. Ron Reagan, Jr. calling on people to support stem cell research, *PBS NewsHour* (July 27, 2004).
31. K. Kamenova and T. Caulfield, Stem cell hype: Media portrayal of therapy translation, *Science Translational Medicine* 7, March 11, 2015: 1–4.
32. Ibid., p. 280.
33. Ibid., p. 278.
34. Craig O. Stewart et al., Beliefs about science and news frames in audience evaluations of embryonic and adult stem cell research, *Science Communication* 30 (June 2009): 427–452.
35. Ibid., pp. 435–436.

36. C. O. Stewart, The influence of news frames and science background on attributions about embryonic and adult stem cell research: Frames as heuristic/biasing cues, *Science Communication* 35, February 2013: 86–114.
37. S. Ho et al., Effects of value predispositions, mass media use, and knowledge of public attitudes toward embryonic stem cell research, *International Journal of Public Opinion Research* 20, May 7, 2008: 171–192.
38. Ibid., pp. 179–180.
39. Ibid., p. 181.
40. M. K. Lizotte, The abortion attitudes paradox: Model specification and gender differences, *Journal of Women, Politics & Policy* 36, February 4, 2015: 22–42.
41. H. Asher, *Polling and the Public: What Every Citizen Should Know*, Washington, DC: Congressional Quarterly Press, 2007.
42. Gallup Poll Social Series are performed during the same month every year and include a number of core questions as well as more topical questions based on current issues. By using the same wording of questions, researchers can analyze the trend dates more reliably. See: www.gallup.com/178685/methodology-center.aspx.
43. J. Wilke and L. Saad, Older Americans' moral attitudes changing, Gallup (June 3, 2013); www.gallup.com/poll/162881/older-american-moral-attitudes-changing.aspx.
44. A Likert or Likert-type scale, named after psychologist Rensis Likert, is an ordered scale used to measure a person's attitudes toward a person, topic, or idea. Responses might be: strongly agree, agree, neutral (neither agree nor disagree), disagree, and strongly disagree. Some scales can extend the categories to include other variations of agree (for example, somewhat agree or mildly agree). These Likert-type response categories give respondents more choices than a simple binary option. If a researcher is interested in how strongly the respondent holds his/her belief, then more options are useful. See G. M. Sullivan and A. R. Artino, Jr., Analyzing and interpreting data from Likert-type scales, *Journal of Graduate Medical Education* 5, December 2013: 541–542.
45. A. Clymer, The nation: Wrong number; the unbearable lightness of public opinion polls, *New York Times*, July 22, 2001.
46. Bernard Roshco as quoted by Adam Clymer, ibid.
47. M. C. Nisbet, The polls—Trends: Public opinion about stem cell research and human cloning, *Public Opinion Quarterly* 68, Spring 2004: 132–155; M. C. Nisbet and A. B. Becker, The polls—Trends: Public opinion about stem cell research, *Public Opinion Quarterly* 78, Winter 2014: 1003–1022.

5 Costs and consequences

Funding fragmentation

> Medical miracles do not happen simply by accident. They result from painstaking and costly research, from years of lonely trial and error, much of which never bears fruit, and from a government willing to support that work. When government fails to make these investments, opportunities are missed. Promising avenues go unexplored. Some of our best scientists leave for other countries that will sponsor their work. And those countries may surge ahead of ours in the advances that transform our lives.
>
> —President Barack Obama, March 9, 2009[1]

The introductory quote for this chapter, taken from remarks made by President Barack Obama after he signed an Executive Order expanding federal funding for human embryonic stem cell research, affirms the consequences, both medical and economic, of not fully supporting scientific research.

Scientific discoveries do not happen in a vacuum, nor do they happen without significant financial support. It can take decades for medical breakthroughs to reach patient therapies. Some findings never bear fruit. These are the inevitable consequences of medical research. A drug that cures cancer in a mouse may not produce the same result in a human. A stem cell therapy that enables a mouse to walk again may not do the same in a human. The costs of clinical trials are not measured in dollars alone, but in the hopes and disappointments of patients as well as researchers. The emotional and political costs of rising expectations have extended the stem cell debate. In signing his Executive Order, President Obama recalled the words Christopher Reeve once told a reporter, "if you come back in ten years. I expect that I'd walk to the door to greet you."[2] Reeve passed away in 2004. The president concluded his remarks by noting "that if we pursue this research, maybe one day, maybe not in our lifetime, or even in our children's lifetime—but maybe one day, others like him might."[3] Was the president contributing to the hope? Of course. We all hope that stem cell research will eliminate words like "terminal" or "incurable" in our lifetime. Cures begin with the research of discovery then move, often slowly, through the various stages of trial to produce a drug or treatment that will cure disease. It is a slow, deliberative, and lengthy process. In 1971, President Richard Nixon signed the National Cancer Act, declaring a war on cancer.[4] We have come a long way in treating cancer, but it continues to be the second leading cause of death in the United States.[5] In June 2016, President Obama confirmed his commitment to cancer research when he launched the Genomic Data Commons located at the National Cancer Institute, "the first-of-its kind, open-access cancer database to allow researchers to better understand the disease and develop more effective treatments."[6] Today, the National Institutes of Health's (NIH's) annual budget for cancer research is more than $6 billion. Stem cell research requires an equally strong commitment and sustained funding. President Obama continued that commitment: funding has increased, and treatments are anticipated.

The need for significant, sustained funding has driven scientists and the media to tout the expectation that therapies are around the corner. It is, perhaps, no surprise that both Dr. James

Thomson and Dr. John Gearhart listed specific conditions that their research would eventually be able to treat. Linking their research to specific conditions such as Alzheimer's disease, Parkinson's disease, and diabetes helps the public visualize the desired outcome.[7] These are no longer undifferentiated cells in a Petri dish, but rather solutions for mediating debilitating and costly conditions. The prospect of a clear mind and a steady hand are worthy investments. These expectations have led to a technology-forcing strategy by funding agencies. Translation has become the operative term. Getting the research from the lab to the clinic and beyond is a common component of current thinking incorporated into the strategic plans of funding agencies.[8]

In this chapter we examine the NIH's role in funding basic research and, more specifically, stem cell research. While other federal agencies fund the research of discovery, none have an annual budget equal to that of the NIH. President Obama's commitment to increasing funding for stem cell research coincided first with the U.S. recession (2009–2010) and second with sequestration (2013). Ironically, the recession led to an increase in NIH funding, resulting from the American Recovery and Reinvestment Act.[9] The Act, more commonly referred to as the Stimulus or the Recovery Act, provided additional funding to government agencies, including the NIH. It may seem odd that some government agencies would see budget increases in the middle of a recession. However, the funds were to foster economic growth by putting more people back to work. The NIH awarded many grants to research labs during this period—providing employment to more scientists, graduate students, and technicians. Sequestration, on the other hand, occurred in 2013 and led to automatic spending cuts for all federal agencies. When Congress passed the Budget Control Act of 2011, it led to across-the-board funding cuts in all government programs for fiscal year 2013.[10] Awarding additional research grants is always more desirable than having to cut funding from ongoing, established projects.

Finally, some state governments (as discussed in Chapter 3) were committed to promoting stem cell research not only to advance the science, but also to expand their economies. Research labs employ a large number of highly educated and well paid people such as scientists, postdoctoral students, and technicians. These labs also attract privately owned biotechnology companies that benefit and assist in the development of clinical therapies. California is home to many privately owned companies. Geron, the company that funded the research of Thomson and Gearheart and initiated the first clinical trial on spinal cord injuries using human embryonic stem cells (hESCs), is located in California. Asterias Biotherapeutics, which resumed clinical trials on spinal cord injuries, is also California-based. ViaCyte, located in San Diego, is running a clinical trial for Type I diabetes that uses hESCs. But states also fell victim to the economic recession that caused early promises of funding to evaporate under the strains of a shrinking tax base.

What role did private funding play in the advancement of stem cell research? While private sector data is harder to obtain, some writers argue that governments (both federal and state) should reduce their spending and let the private sector fund the research. These authors counter the argument made by scientists that the private sector would not support research of discovery by noting that companies such as Advanced Cell Technologies, Stemcells, Inc., and ViaCell, along with philanthropists like Michael Bloomberg (he donated $100 million to Johns Hopkins Medical School for stem cell research), did and will invest in this medical field.[11] Then why waste public money when the private sector has a good track record of moving discoveries from laboratory to clinic? An example of a medical technology that developed without any public funding is in vitro fertilization (IVF) research. This example is often given as evidence that science can rely on private funding. This may be true, but discoveries are often made by scientists whose initial research was supported with public funds (see Chapter 2 for a discussion of IVF).

The second part of this chapter discusses the nonmonetary consequences of stem cell research. In this context, perception is as important as reality; the loss of national dominance over an emerging technology is one of the more difficult consequences to measure. The prestige of the United States as a global leader in biotechnology, while difficult to measure in quantitative terms, was being jeopardized as a direct consequence of limited funding for hESC research.[12] Two American scientists, James Thomson and John Gearhart, had succeeded in isolating hESCs, something researchers across the globe had also been trying to accomplish. Scientists and entrepreneurs were concerned that the United States would relinquish the opportunity to develop these cells into medical therapies. A related consequence, more perceived than real, was the brain drain of scientists leaving the United States for countries with more favorable policies toward hESC research.[13] In 2004, the perception of an internal brain drain emerged after California passed Proposition 71: The California Stem Cell Research and Cures Initiative (Prop. 71). Some feared that the siren call of California's money would cause scientists to leave their labs for the Golden State and its dollars.[14]

The NIH: Funding the research of discovery

Who funds basic research? The International Council for Science defines basic research as "fundamental theoretical or experimental investigative research to advance knowledge without a specifically envisaged or immediately practical application. It is the quest for new knowledge and the exploration of the unknown."[15] While basic research may produce no immediate products or cures, it carries the expectation that over time some of the research will lead to medical benefits.

In a comprehensive report on the benefits of the NIH, published by the U.S. Senate Joint Economic Committee, Senator Connie Mack (R-FL), chair of the Committee, noted that "the benefits derived from our commitment to research have led to life-saving medical breakthroughs and dramatically improved the quality of life for people everywhere. We are on the verge of finding cures and new treatments for so many of the diseases that plague our society. Research is the key to unlocking the knowledge we need to find these cures. Investment in basic science helps us compete in the global marketplace in such industries as pharmacology, biotechnology, and medical technologies."[16] Similarly, President Obama also noted in his remarks on expanding funding for stem cell research that "from life-saving vaccines to pioneering cancer treatments, to the sequencing of the human genome—that is the story of scientific progress in America."[17] This is a resounding endorsement that NIH research leads to tangible benefits. Maybe not immediately, but certainly over time. The Centers for Disease Control and Prevention (CDC) confirms that life expectancy at birth in the United States increased from 47 years in 1900 to 78.8 years in 2014.[18] The top three causes of death in 1900 were influenza, tuberculosis, and gastrointestinal infections. Today, these illnesses are rarely fatal.[19] The development of a vaccine to combat the flu in the early 1980s, which was safe enough for even infants, resulted from NIH-funded research. While cancer is the second leading cause of death in the United States today, death rates from cancer have fallen significantly, due in part to NIH-supported efforts. According to the American Cancer Society, NIH-funded research "has played a role in every major cancer prevention, detection, and treatment advance" over the past several decades.[20]

When hESCs were isolated by Thomson and Gearhart in 1998, there were no immediate applied benefits. Although both researchers speculated that eventually their discoveries would lead to treatments and cures for many diseases, it was not until 2009 that the first clinical trial using hESCs was approved by the U.S. Food and Drug Administration (FDA). The privately held Geron Corporation initiated, and later suspended, a trial using hESCs to treat spinal cord injuries. This trial has subsequently been taken up by another private company, Asterias Biotherapeutics, and was enrolling patients at the time of writing.[21] Clinical trials for Alzheimer's and Parkinson's treatments mentioned by Thomson have yet to emerge. However, clinical trials for Type I diabetes, approved by the FDA in 2014, began enrolling participants in 2016.[22] Moving from the lab to the clinic is a long process. As one scientist working on hESC research noted: "These papers represent seven or eight years of work."[23] Seven or eight years might sound like a long time, but in scientific time it is a relatively short time frame, especially considering that hESCs were first isolated in 1998. Researchers had not yet figured out how to get these pluripotent cells to differentiate into all of the cells that make up the human body. The field has matured as scientists working in labs across the globe share their findings and build on new discoveries. The research of discovery is moving into the research of delivery. Clinical trials are the first step. Successful trials usually encourage private companies to move into the area. As we noted in Chapter 2, Pfizer, a global biopharmaceutical company headquartered in New York, is funding the clinical trials on age-related macular degeneration. The pharmaceutical company Johnson & Johnson is contributing financially to the ViaCyte trial on Type I diabetes. The science of delivery requires that partnerships be established between researchers and biomedical providers. As we will discuss later in this chapter, NIH and the California Institute for Regenerative Medicine (CIRM) have clearly signaled that the next step of moving treatments to patients will include partnerships with private enterprise.

In the United States, the major agency funding basic research is the National Institutes of Health (NIH). Originally called the National Institute of Health in 1930 when it was first created, its name was changed in 1944 when the National Cancer Institute (NCI) became a division of the NIH. Today, there are 27 separate institutes and centers that make up the NIH campus located in Bethesda, Maryland. Each of the centers has its own research agenda that is often clearly linked to a specific disease, such as the National Cancer Institute or the National Eye Institute. But some of the institutes, such as the National Institute of General Medical Sciences (NIGMS), established in 1962, support basic research, including diseases or conditions that are the focus of the other institutes.[24] Colocated on the Bethesda campus is the NIH Clinical Center, where many of the clinical trials are carried out by both internal NIH researchers (employees of the federal government) and external, NIH-funded researchers (employees of public and private universities or private companies).[25] In 2012, the National Center for Advancing Translational Sciences (NCATS) was established. The creation of NCATS affirms the importance that the NIH places on getting treatments and cures to patients faster; the center emphasizes innovation and deliverables.[26] As evidence that the private sector plays an integral role in translation, in 2014 NCATS announced a collaboration with Pfizer's Centers for Therapeutic Innovation. Pfizer established similar collaborations with academic institutions and patient foundations. Pfizer's four translation hubs are located in Boston, New York, San Diego, and San Francisco. It is not a surprise that two of the centers are in California, the state that financially promoted hESC research. Pfizer puts its scientists in labs with academic investigators, providing funding and access to proprietary data. In exchange, Pfizer offers "equitable intellectual property and ownership rights."[27] Medical products and procedures developed by scientists are inventions and can be patented. Ownership of the product (drug or device) or the procedure (how to derive stem cells or administer a treatment), if patented, requires a payment or some other form of acknowledgement. In 1998, James Thomson was issued two patents for his hESC research, both of which were owned by the Wisconsin Alumni Research Foundation (WARF) of Madison, Wisconsin. The foundation owned the stem cell lines and made them available for researchers around the world for a price ($5000 inside the United States and $6000 outside the United States).[28] The market for translation is increasing with the help of private companies. In exchange for financial support, these companies seek the ownership rights that allow them to develop and market the products or procedures. This arrangement is welcomed by research facilities, as most of them are not equipped to produce a drug or treatment for the mass medical market. Once a treatment has been proven effective (following clinical trials), scientists are eager to see it used by patients. This validates the significance of their work in curing diseases.

The NIH is the main source of federally funded basic research. There are, however, other federal agencies that also fund some basic research, the most well-known (though not exclusively medical) being the National Science Foundation (NSF), which was created by Congress as an independent agency in 1950 "to promote the progress of science, to advance national health, prosperity, and welfare...."[29] Statistics cited by the NSF indicate that they support 200,000 scientists, engineers, educators, and doctoral students across the 50 states. The NIH provides similar data, indicating that "more than 80% of the NIH budget is awarded through about 50,000 competitive grants to more than 300,000 researchers at more than 2500 universities, medical schools and other research institutions in every state and round the world."[30] Unlike the NSF, which does not conduct in-house research (all of its money is awarded externally), the NIH budget supports 6000 in-house scientists working in labs at the institutes and centers that make up the NIH.

Other federal agencies that engage in basic research include the Veterans Affairs Medical and Prosthesis Research Program, the U.S. Department of Energy Office of Science, and the U.S. Department of Agriculture's Food Research Initiative. However, funding amounts vary considerably. While the NIH's fiscal year (FY) 2016 budget was $32 billion and the NSF budget was $7.46 billion, the Department of Energy Office of Science had a budget of $5.35 billion, the Department of Agriculture Research Service, $1.14 billion, and the Veterans Affairs Medical and Prosthetic Research Program, $630.7 million.[31] The Department of Defense also funds basic research but primarily for military hardware; these figures are not available for public review.

Table 5.1 provides an overview of NIH and NSF total funding from 2002 to 2017. We calculated that the number of NSF grants supporting stem cell research in fiscal year 2013 was 392, compared with more than 8000 for the same time period by the NIH.

The average annualized award size for the NSF in FY 2013 was $161,200. These grants may be renewed for up to 3 years.[32] On the other hand, NIH's average annual grant award size was $441,404 for FY 2013.[33] The NIH grants may be renewed for up to 5 years. While the differences in grant support for basic research may seem staggering, a report posted in January 2014 by the head of extramural research, Dr. Sally Rockey, notes that there has been a decrease in the average award size in NIH research grants from $454,588 in FY 2012 to $441,404 in FY 2013.[34] An even more telling figure is the decline in what that money can purchase (the inflation factor). Dr. Rockey notes that the average size of a research grant in 1999 declined from

Table 5.1 NIH and NSF total funding 2002–2017 (in millions of dollars)

	NIH		*NSF*	
2002	23,188		4825	
2003	26,740		5323	
2004	28,100		5588	
2005	28,626		5482	
2006	28,533		5589	
2007	29,034		5889	
2008	29,320		6125	
2009	30,207	10,400[a]	6493	3002[a]
2010	31,036		6872	
2011	30,630		6805	
2012	30,802		7033	
2013	28,137[b]		6884[b]	
2014	30,019		7171	
2015	30,294		7344	
2016	32,311		7463	
2017	33,136		7964	

Source: NIH budget. NSF budget.

Notes

a American Recovery & Reinvestment Act Funds FY 2009.

b Budget reductions due to sequestration.

$290,869 to $277,653.[35] Scientists are being asked to do more with less money. She also noted that the number of successful research project grants declined from 17.6% to 16.8% between FY 2012 and FY 2014.

When you consider the fact that more than 80% of the scientists applying for NIH research support are not successful, you better understand the problem faced by promising researchers who seek to continue their work. In a polemical piece, Kwame Boadi, from the Center for American Progress, writes that the "funding figures obscure the critical factor that affects America's capacity to continue leading the world in biomedical research—inflation. The cost of conducting biomedical research is increasing faster than the cost of other goods and services in the economy.... Aside from the very real impact that this reduction in grants is having on the lives of researchers who are unable to pursue their livelihoods, the impact on society at large will not be felt to a large extent either today or tomorrow, but it will absolutely be felt over the next decade or two. In order to put NIH funding back on track, significant new investments must be made in the coming years to cement the federal government's commitment to biomedical innovation."[36] A 2014 study by the American Academy of Arts and Sciences concluded that Congress needs to restore research funding if the United States is to retain its position in the world.[37]

Despite the limitations on federal funding during the Bush Administration, the interest in stem cell research exploded. Students in the United States and abroad rushed to complete degrees in cell biology and related fields. It became clear that there were not enough scientists trained in how to isolate and maintain hESCs. A review of the NIH budget for FY 2002 and 2003 indicates that some of the funding was to train researchers on how to culture hESCs and workshops on stem cell protocols.[38] The 2013 budget cuts, resulting from sequestration, further reduced available science funding. A study conducted by the American Society for Biochemical and Molecular Biology (the Society) affirmed that federal funding for science had lost about 20% of its purchasing power over the last 3 years.[39] Responses to an online questionnaire administered by the Society illustrate the dilemma of too many scientists chasing too few dollars. A graduate student from California responded: "It is disheartening to be at the start of what I hope will be a strong and successful scientific career and have to wonder if I will even get a job, be able to fund my research and have hope of being a competitive scientist."[40] Similarly, a postdoctoral scholar lamented: "I am amongst a growing list of scientists that the federal government has spent $200,000 to $400,000 to educate and train since the early 1990s. The expectation was that we would have the opportunity to show U.S. taxpayers a return on their investment by becoming (science, technology, engineering, and mathematics) leaders and innovators far into the middle of the twenty-first century. Unfortunately, the current funding climate presents the real possibility that the taxpayers may witness a significant loss in their investment."[41] These comments provide a glimpse into the competitive world of scientific research. The research of discovery has long depended on federal funding. When funds are cut because the research is controversial (as during the Bush Administration) or because revenue declined (sequestration), scientific research is put on hold or moved to another country. Either way, results are delayed. This scenario might explain why more clinical trials using hESCs have not emerged. Sound science takes time and money.

Funding for hESCs

Now that you have an idea of what a big player the NIH is in the biomedical field, it puts the tension between President George W. Bush and researchers in sharper perspective. Most researchers in 2001, as today, look to the federal government to fund their work. Since NIH grants may be awarded to private for-profit research institutions, university researchers are also competing with scientists working at Johnson & Johnson and other large pharmaceutical facilities. The bottom line is that the NIH is interested in finding and funding cures; a lot of work is done by private companies.

How much did the NIH fund? After President Bush's speech limiting hESC funding to only those lines derived prior to August 9, 2001, the NIH had to issue new grant guidelines, which it did on November 7, 2001.[42] The first grants to stem cell scientists were not awarded until FY 2002 (the federal fiscal year begins October 1 and ends September 30).

In Table 5.2 we look at NIH funding from 2002 through 2017 to provide an index of NIH support for stem cell research in general and hESC rsearch in particular. It was clear from President Bush's statement in August 2001 that he was conflicted about the research but was moved by the scientific community as well as advocacy groups, both of which were hoping for quick cures. In Chapter 5 we noted that, based on polls and surveys done at the time, public attitudes were generally in favor of the hESC research. However, very little research using hESCs had been done on specific conditions. The first clinical trial began in 2009. Several other trials using hESCs were approved in 2014 (see Chapter 2).

In this table (Table 5.2) we also include funding figures for four of the leading causes of death and disability in the United States—heart disease, cancer, Alzheimer's disease, and diabetes. Heart disease and cancer have been first and second on the list of the 10 leading causes of death since 1990. Diabetes is the sixth or seventh leading cause of death (depending on the year). Alzheimer's was included on the list of the 10 leading causes in 2000.[43] According to the Alzheimer's Association, in 2016 Alzheimer's became the sixth leading cause of death.[44] The cost of caring for individuals with Alzheimer's was estimated to be $236 billion in 2016.[45] Finding a cure for these leading causes of death and disability is a national medical priority.

The NIH provides an extensive database of how it allocates the billions of dollars it receives annually. The passage of the National Institutes of Health Reform Act of 2006 (P.L. 109-482)[46] mandated the creation of an electronic database—the Research, Condition, and Disease Categorization (RCDC) system. The electronic system can be searched by diseases, conditions, and funding and the database, designed to be fully transparent, can be searched by principal investigator, university, or state. Grant applicants can find out who received funding and read the research abstracts. Members of Congress who want to know how much funding was received in their congressional district can search by this variable. The Federation of American Societies for Experimental Biology (FASEB) annually constructs an interactive map of the United States broken down by congressional district that indicates how many research dollars flowed into each district.[47]

In 2009, the RCDC database listed 233 funding categories; there are currently 265 categories. The list includes actual diseases such as diabetes and cancer, conditions such as aging, and research such as hESC.[48] In 2012, the RCDC included funding figures for induced pluripotent stem cells (iPSCs). These stem cells were derived in 2006, and research funding prior to 2012 was included in the category of human nonembryonic.

The list is quite detailed, and, combined with the NIH Research Portfolio Online Reporting Tools (RePORT) system, it allows interested individuals to click on each disease or category and gain information on funded projects. Since the system is an active database, NIH staff members remind individuals that numbers can change as projects are completed.[49] The NIH funding can be extended for up to 5 years, so it is common for additional funding requests to change the original figures. We are interested in how much money was spent on human embryonic stem cell research compared with other types of stem cell research, and stem cell research with other major disease categories.

Table 5.2 reveals that cancer research consistently received the most funding between 2002 and 2017. Between 5% and 6% of the NIH budget is spent on cancer research. Heart disease, which is the number one cause of death, received between .03% and .05%. However, NIH staff point out that since heart disease has many contributing factors, research in other institutes might also be funding heart-related research.[50] Examining the figures for NIH-funded human stem cell research shows that funding between 2002 and 2007 looks very anemic. The numbers double in FY 2008 to $88 million, and $166 million in FY 2010. But compared to how much was spent during the same time period on non-human hESC ($72 million in FY 2002 and

Table 5.2 NIH research funding 2002–2017 for major disease categories and stem cell research (in millions of dollars)

	2002	*2003*	*2004*	*2005*	*2006*	*2007*	*2008*	*2009*[a]	*2010*[a]	*2011*	*2012*	*2013*	*2014*	*2015*	*2016*	*2017*
Heart Disease	598	626	633	703	783	2205	2686	1953	2101	1673	1746	1634	1645	1688	1758	1758
Cancer	3774	4015	4200	4496	4730	6105	6278	6749	6626	5440	5621	5274	5392	5389	5652	6332
Alzheimer's	57	44	192	160	161	685	722	543	529	448	503	504	562	589	910	910
Diabetes	1137	1125	1874	1727	1617	1123	1416	1151	1199	1076	1061	1007	1011	1010	1044	1044
Stem Cell Research	387	517	552	609	643	658	1032	1231	1286	1179	1374	1273	1391	1429	1495	1495
hESC Human	10	20	24	40	38	42	88	143	166	123	146	146	166	180	191	191
hESC Non-Human	72	114	89	97	110	106	150	177	195	165	164	154	150	159	168	168
Non-hESC Human	171	191	203	199	206	204	297	397	415	395	504	431	443	445	465	465
Non-hESC Non-Human	134	192	236	273	289	306	497	638	644	620	653	613	627	632	658	658
iPSC											206	228	313	324	339	339
iPSC Human											175	199	280	282	296	296
IPSC Non-Human											48	43	49	61	63	63
Total NIH Funding	23,188	26,740	28,100	28,626	28,533	29,034	29,320	30,207	31,036	30,630	30,802	28,137	30,019	30,294	32,311	33,136

Source: Calculated from NIH RePORT database.

Note
a Includes American Recovery & Investment Act funding.

$195 million in FY 2010), the figures for hESC research are meager. This low number can be explained, in part, by the fact that the hESC research field was in its infancy; scientists did not know how to work with these new cells. We used the NIH RePORT system to look at what was being funded during these early years. Our research revealed that funds were allocated in three general categories: (1) the expansion and distribution of hESC lines; (2) training for prospective researchers; and (3) research grants to scientists actually doing research with stem cells.[51]

President Bush permitted research on those hESC lines derived prior to his August 9, 2001 speech. At the time there were 78 approved lines, but only 22 were available.[52] Some of the research centers that owned approved lines (such as the Geron Company or the University of Gothenburg in Sweden) did not offer their products for sale. Of the 22 available lines, four were owned by research groups outside the United States. E.S. Cell International, an Australian company, owned six lines, Technion University in Israel owned three lines, and Cellaritis AB in Sweden owned two lines. In the United States, the WiCell Institute, an affiliate of the Wisconsin Alumni Research Foundation, owned five lines, and BresaGen, Inc., a private research company in Georgia, owned three lines.

When we looked at funding for hESC research from FY 2002 to FY 2005, we found that the NIH funded several of the private companies and research centers that derived hESCs, providing them with funds for the expansion and distribution of the approved lines. BresaGen, Inc. received $1.6 million to make vials of hESCs available to researchers. The company, as did others that owned hESCs, charged researchers $5000 per vial, a number recommended by the NIH.[53] E.S. Cell International received almost $1 million in FY 2002 and another $500,000 in FY 2003. Technion University received about $500,000 in both years for preparation and distribution of hESC lines. The WiCell Institute received $550,000 to prepare the lines they owned. Between FY 2005 and FY 2010, WiCell was designated by the NIH as the National Stem Cell Bank. In FY 2010, when NIH funding ended, the stem cell bank was absorbed by WiCell and became known as the Wisconsin International Stem Cell Bank. But by 2010, many public and private organizations had derived their own stem cell lines and were no longer restricted to using the federally approved lines. Some researchers, like Doug Melton from Harvard's Stem Cell Institute, used private funds to derive their own hESCs as early as 2002. Melton reportedly exclaimed: "I can't wait to give them out!"[54] Melton's hESCs were not on the approved list, so researchers could work with the cells, but they could not apply for federal funding.[55] This changed in 2009, when President Obama expanded federal funding, adding additional stem cell lines to the registry approved for federal funding.

Beginning in FY 2003, the NIH established three extramural exploratory centers for human embryonic stem cell research. Each center was funded for a 2-year period and had a different specific mission. Total NIH funding for the centers was $6.3 million.[56]

The three centers were:

1. The University of Washington, Seattle/Fred Hutchinson Cancer Research Center, $753,000 to "improve methods to culture, maintain, manipulate, differentiate and compare the 12 federally approved embryonic stem cell lines."[57]
2. The University of Michigan Medical School in Ann Arbor received funding ($778,000) to "apply knowledge and expertise in cell biology, developmental genetics and tissue biology."[58] The University was also supposed to support the education and training of stem cell researchers.
3. WiCell Research Institute in Madison, Wisconsin received funding ($669,000) to "create a central core facility to provide cell tissue culture support, including media preparation, quality control and routine chromosomal analysis of cultured stem cells."[59] Since this is the home campus of James Thomson, he became the principal investigator for this center.

You might be thinking that NIH funding was designed to jump-start hESC research. In fact, that is correct. According to the director of the NIH, Elias A. Zerhouni, "[t]here is so much basic research we must conduct before we can unlock the potential of these cells and fulfill their promise."[60] The hope that these cells would be ready for human trials in a few years proved to be overly optimistic. The NIH recognized that there was a shortage of trained researchers and was allocating money to create an educated and skilled scientific workforce. Since the NIH could not use any money to derive new stem cell lines, researchers had to work with the approved lines in the most effective way possible.

This leads us to the second most frequent (though not most costly) NIH expenditure during this early period—funding for stem cell workshops and training sessions.[61] Starting in FY 2003 and continuing through FY 2016, the NIH funded numerous workshops, short courses, and how-to sessions. These short courses were taught by top stem cell researchers. The courses taught between FY 2003 and FY 2005 tended to be of a more general nature, at least for the stem cell research community. In FY 2003, three of the courses taught included: Current Protocols in Stem Cell Biology, Frontiers in Human Embryonic Stem Cell Research, and Short Course in Human Embryonic Stem Cell Culture; you do not need a science degree to figure out what might be on the agenda. Together the cost for the three courses was less than $300,000. The stem cell culture course was a popular one during this early period as it was repeated with some frequency until FY 2008.

One of the early issues with hESCs was that most of the approved lines had been grown on mouse feeder cells. For early researchers trying to figure out if replating these existing lines might be improved by changing the culture material was an important task. By FY 2010, when there were more scientists working with hESCs, the workshops became more technical and more costly. For example, of the nine workshops funded by the NIH in FY 2010, one called Integrated Technology Resources for Biomedical Glycomics, was funded for $1,305,160. The course Frontiers of Human Embryonic Stem Cells was updated, renamed Frontiers in Human Embryonic Stem Cells and Regeneration, and given more funding. These courses were taught in the United States and in other countries by both private corporations and public institutions. Two of the private companies that owned approved hESC lines, BresaGen, Inc. and E.S. Cell International, had employees teaching courses on the efficient distribution of hESCs. The NIH continues to fund stem cell training sessions; in FY 2015 the NIH funded 15 workshops/training sessions, but the topics were even more specific and targeted to specific diseases. So the course Frontiers in Human Embryonic Stem Cells and Regeneration became Frontiers in Stem Cells in Cancer. It continues to be taught by Professor Gerald Schatten from the University of Pittsburgh School of Medicine, a leading scientist in the area of stem cell research and molecular medical therapies.

A number of workshops focus on the training of new and existing scientists. For example, some of the 16 courses funded by the NIH in 2015 include a predoctoral training program in genetics, a graduate training course in cellular and molecular pathogenesis of human diseases, and a medical scientist training program. A second common theme is translation. The focus is not just knowledge and information, but moving the research out of the lab and into the clinic. These courses include a training program in stem cell translational medicine for neurological disorders and kidney development cell fate and precursors of disease in the young and adult. The NIH's new mission is getting the research out into the medical arena.

The third focus of NIH is funding research grants. The number of funded hESC grants increased from 54 in FY 2002 to 1862 in FY 2015. The numbers were very low in the early years for some of the reasons discussed previously. If you look at the researchers who received funding in FY 2002, you find the names of the leading scientists who were first to work in the field. John Gearhart, one of the scientists credited with isolating hESCs, received funding for a study looking at stem cells and Down syndrome. George Q. Daley, from the Whitehead Institute for Biomedical Research and later Harvard's Stem Cell Institute, received two grants for work comparing hematopoietic stem cells with hESCs. Doug Melton received funding for his research on the genetic regulation of hESCs, and, of course, James Thomson received funding for his research on self-renewal of primate hESCs. The institutions that received funding during this early period represented large, established scientific centers: Stanford University, Harvard University, University of California, San Francisco, and the University of Michigan, to name but a few. By 2015, the number of research grants had increased, as had the number of different affiliations of researchers. The same big 10 or 20 institutes continue to be well represented in the funding stream, but smaller centers also appear to be competitive, such as Smith College, University of Denver, and the University of Illinois at Chicago and University of Illinois at Urbana-Champaign. Affiliations now also include clinical hospitals and labs such as Sloan Kettering Institute for Cancer Research, Mayo Clinic, and Children's Hospital of Philadelphia.

The field matured as more researchers started working with hESC then expanded as scientists moved to apply the research to existing diseases. The key for the later period is translation. How can we move beyond the lab and into the clinic? In the next section we look at how state governments, especially California, took up the same challenge.

The NIH was not the only game in town. When President Bush limited funding, as noted in Chapter 3, a number of states moved to fill the gap. By way of reminder, after the isolation of hESCs in 1998, the NIH published guidelines for federal funding of research using these cells in August 2000. An NIH panel was scheduled to evaluate the proposals in April 2001 (only three proposals were received); the meeting was postponed to give the president time to review the issue.[62] Thus, in effect, until FY 2002, no federal money had been allocated for hESC research.

Between 1998 and August 2001, stem cell scientists as well as their universities and research institutes were concerned that the federal government would not fund any research using hESCs. Some feared that a law might be passed (some states did just that) prohibiting any research using hESCs. As a reminder, there is no federal law prohibiting the derivation of hESC lines; the prohibition is on the use of federal funds to derive these lines. This prohibition still exists. Individual states, however, were not prevented from funding this research; nor was the private sector.

The National Conference of State Legislatures (NCSL) compiles and regularly updates statistics comparing state laws in a number of policy areas. One such compilation includes the status of state laws regarding research on fetuses and embryos.[63] Table 5.3 is adapted from the data reported by the NCSL to reflect only those states that specifically permit or prohibit research using embryonic stem cells. It bears reiterating that the federal government does not have a law prohibiting the derivation of hESC or even therapeutic cloning. Federal funds, however, cannot be used for these activities.

Table 5.3 reveals that relatively few states have specific laws permitting stem cell research. Most of these states were discussed in Chapter 2 as being among the first to financially support hESC research. Michigan entered the arena later than other states, amending its constitution in 2008 to permit a "full range of stem cell research."[64] We mentioned earlier that the University of Michigan Medical School was one of the NIH's designated exploratory centers. Funding for the center was provided by the NIH, and only approved hESC lines could be used. The 2008 law cleared the way for researchers in Michigan to derive new lines with private funding. The ballot initiative approved in Michigan contains the following language: "No stem cells may be taken from a human embryo more than fourteen days after cell division begins; provided, however, that time during which an embryo is frozen does not count against this fourteen day limit."[65] This language should sound familiar, as it is strikingly similar to that found in England's Warnock Commission Report.[66]

A number of states are not listed as either permitting or prohibiting hESC research. Does this mean that no research is going on? Probably not. We say this because at least one of the private companies on President Bush's approved hESC registry in 2001 was BresaGen, which is located in Georgia, a state that does not appear on the table. Private companies were permitted to spend their own money on hESC research, including the derivation of new stem cell lines. The states that spent their own money pushing this research became major players. This is affirmed, in part, by several studies. One study looked at "the geographic preferences of

Table 5.3 States with specific laws on human embryonic stem cell research

	Permit	*Prohibit*
Arizona		X
Arkansas		X
California	X*	
Connecticut	X*	
Florida		X
Illinois	X*	
Indiana	X*	
Iowa	X	
Kentucky		X
Louisiana		X
Maine		X
Massachusetts	X*	
Michigan	X*	
Minnesota		X
Missouri		X
Montana		X
Nebraska		X
New Hampshire		X
New Jersey	X*	
New Mexico		X
New York	X	
North Dakota		X
Ohio		X
Oklahoma		X
Pennsylvania		X
Rhode Island		X
Tennessee		X
Texas		X
Utah		X
Virginia		X
Wyoming		X

Source: National Conference of State Legislatures 2015.

Notes
States not listed do not have laws either permitting or prohibiting research.
X* indicates consent needed to donate embryo and conduct research.

scientists" and found that supportive state policies (like those in California) were a strong factor in influencing relocation decisions by stem cell scientists.[67] Another study concluded that state funding in California and Connecticut contributed to an "over-performance" of scientific publications on hESCs.[68] Publication rates are one indicator, but given the size and funding advantage of California over other supportive states, the finding is not surprising. Beginning in 2007, California was spending $300 million per year compared with between $10 and $55 million in other states.[69]

Although a number of states passed laws permitting hESC research, most states were not able to sustain their financial commitment. New Jersey provided early seed money for the establishment of a stem cell institute at Rutgers University, but when public funding was rejected by voters in 2007, the institute was put on hold. Wise Young, a leading stem cell researcher and active promoter of the New Jersey project, continues to be active in the field.

He is currently conducting clinical trials but using umbilical cord stem cells.[70] Similarly, Massachusetts ran out of public money to support its stem cell bank and registry in 2012.[71] The bank was created in 2008 when new hESC lines were still not eligible for federal funding. This all changed in 2009 when President Obama expanded funding to include many of the newly derived hESC lines. Despite an increase in federal funding, some states continue to maintain stem cell banks, such as WiCell in Wisconsin. In 2013, California invested $16 million to create three new stem cell repositories—one for tissue collection for disease modeling, one for iPSC lines, and one for hESC lines.7[2] The stated objective was to generate and ensure the availability of a high quality, disease-specific line.[73] All three grants were awarded to Cellular Dynamics International, Inc. (located in Madison, Wisconsin), the same company that manages WiCell's stem cell bank.[74]

In addition to NIH, California became the only major player in hESC research using public funds. Although research was going on all over the world,[75] California put a high priority on public support for biomedical research. The state already had a strong foundation in this research area—world-class universities and research centers, well-known stem cell scientists, and established private biotechnology companies. Alice Park, a respected science journalist, provides an informative review of the factors that led to the passage of Proposition 71: The California Stem Cell Research and Cures Initiative.[76] By way of summary we highlight a few.

Although President Bill Clinton opened the prospects of federal funding for stem cell research as early as 1995,[77] it was not until late 2000 that the NIH began accepting grant proposals for actual funding.[78] One of the proposals that the NIH received was from Roger Pedersen, a stem cell researcher at the University of California, San Francisco (UCSF). Pedersen had also succeeded in isolating his own hESC with private funding from the Geron Corporation of Menlo Park, California (the company that also funded Thomson and Gearhart). When Pedersen learned that NIH was canceling all of the submitted proposals until President Bush could revisit stem cell policy, he decided to move his lab to Cambridge, England. Pedersen said the move afforded him "the possibility of carrying out my research with human embryonic stem cells with public support."[79] Pedersen's announcement months before President Bush's August 9, 2001 speech sent shock waves throughout California's scientific community. An article published in the *Los Angeles Times* summed up the conundrum: "Uncertainty is Thwarting Stem Cell Researchers."[80] Even after the president's speech, uncertainty seemed to prevail. Another UCSF researcher said that he, too, had given serious thought to leaving the United States because of the uncertain legal status of stem cell studies in humans.[81]

In an effort to stem the exodus of the best and the brightest, California passed a law permitting hESC research. Governor Grey Davis signed the bill (SB 253, introduced by State Senator Deborah Ortiz [D]), acknowledging that "with world-class universities, top-flight researchers and a thriving biomedical industry, California is perfectly positioned to be a world leader in this area."[82] The law specifically stated "that research involving the derivation and use of human embryonic stem cells, human embryonic germ cells, and human adult stem cells from any source, including somatic cell nuclear transplantation, shall be permitted…."[83] Somatic cell nuclear transfer is the scientific term for cloning. Laws that support this procedure are more likely to use the scientific term; when the goal is to prohibit the procedure, the word "cloning" is used. For example, the Human Cloning Prohibition Act of 2001 introduced in the U.S. Congress (but never passed) would have prohibited both therapeutic and reproductive cloning.[84]

Although it was assumed that "the legislation would attract the best and brightest of the world's scientists to California,"[85] in fact, the bill was largely symbolic because no money was appropriated to support the research. Scientists were still dependent on private funds or the federal government if they agreed to use the approved hESC lines. Researchers who planned to derive their own hESC lines needed to separate their labs and equipment from any of their other projects that received federal funding. In the interim, several universities received private money to set up stem cell research programs. UCSF received $5 million from Andy Grove, the former CEO of Intel; Stanford University received a $12 million anonymous donation to set up a similar program.[86] Universities in other states were engaged in similar arrangements with private donors to begin research deriving new hESC lines.

Frustration in California was running high when a South Korean team published an article in March 2004 announcing that they had succeeded in creating hESCs from a cloned embryo.[87] The study was later proven to be false, but this was something that scientists in the United States and elsewhere had been trying to accomplish. Creating stem cells from cloned embryos could reduce immune rejection problems when cells are injected into another person.

Newspapers fueled the frustration with articles asserting: "It's official: The United States has fallen far behind in mining the promising field of stem cell research to treat disease."[88] Scientists complained that the discoveries made by U.S. researchers were now being developed outside the United States. Countries like China, England, and Israel were spending more money on stem cell research and, as a result, moving ahead. The government of Singapore was about to open a new $287 million biotechnology center, "Biopolis," to focus on stem cell research.[89]

Finally, a combination of events created a perfect storm that led to the introduction and passage of Prop. 71, an amendment making stem cell research a state constitutional right and backing that right up with $3 billion in state funding. The people, the events, and the money expended to ensure success of the measure could fill volumes. "Depending on whom you ask, you'll get a slightly different version of the story about who came up with the idea of bringing stem cell research to California."[90] Some credit Deborah Ortiz, the state senator who proposed the first law, with getting the ball rolling; others believe that it was Robert Klein, the wealthy real estate developer who chaired the Prop. 71 group. Still others contend that it was the combination of scientists and patient advocates who pressed legislators for increased funding.[91] The death of President Ronald Reagan from Alzheimer's in the summer of 2004 and of Christopher Reeve, an ardent advocate for stem cell research, in fall of the same year galvanized the campaign effort. Ron Reagan's speech (President Reagan's son) at the 2004 Democratic Convention focused the national media spotlight on the California initiative.

Many of the individuals connected with California's push to expand hESC research were directly affected by conditions that hESC research claimed it might cure. Michael J. Fox, a popular Hollywood actor, suffered from Parkinson's and Christopher Reeve had a severed spinal cord. Robert Kline, a wealthy donor, had a son with diabetes. Several other contributors also had family members with debilitating illnesses. There is no doubt that the California campaign pushed the envelope by suggesting that cures were around the corner. The campaign was awash in money with which to convince the voters that everyone in the state would benefit from more research funding, whether it was directly (cures) or indirectly (jobs).

The Coalition for Stem Cell Research and Cures constituted a broad collection of groups that included Nobel laureates, movie stars (Brad Pitt and Michael J. Fox), state elected officials, and many patient and disease advocacy groups. The "Yes on 71 Initiative" created a webpage, still accessible at the University of California, Los Angeles' digital media archives, that included a list of all the coalition's members, contributors, and video and campaign ads. The elements of a successful California initiative can be observed by examining the webpage.[92] The webpage includes links to a section entitled "Stories of Hope," which showcased individuals living with conditions that hESC could cure. This was clearly an effort to put a human face on the research and, in retrospect, to create the expectation of instant cures. There is also a video/audio section that includes some of the ads that were aired in 2004. One ad that was aired posthumously shows Christopher Reeve asking voters to support Prop. 71 (Reeve passed away in October 2004).

The coalition raised $35 million, which it used to campaign throughout the state. Donations came from a combination of corporate leaders, venture capitalists, real estate developers, and Hollywood actors. Robert Klein, the coalition chair, contributed $3 million; other major contributors included Bill Gates (Microsoft), Pierre and Pamela Omidyar (eBay), Gordon Gund (Cleveland Cavaliers), and William Bowes (Amgen). The Juvenile Diabetes Research Fund contributed $1 million.[93] The money was used to run campaign ads that featured Brad Pitt, Michael J. Fox, and Christopher Reeve. One of the ads features Michael J. Fox saying: "Vote yes on 71, and save the life of someone you love."[94] Who could say no?

By comparison, the opposition raised about $400,000 and included the Roman Catholic Church, the California Pro-Life Council, and the Orange County Republican Party. Two groups were created specifically to oppose the initiative: the Pro-Choice Alliance Against Proposition 71 and Doctors, Patients, and Taxpayers for Fiscal Responsibility. The clear pro-life stand of the opposition is evident, but there was also a growing concern with financial accountability.[95] In the end, Prop. 71 was supported by a vote of 59–41%. California is one of 26 initiative states, and it is not uncommon to find anywhere from 10 to 20 initiatives on election ballots.[96] A successful initiative in a large state like California requires a well organized and well funded campaign.

Although the law gave priority to hESC, the funding could in fact be used for other types of stem cells. As we will point out later in this chapter, money was also used to construct new facilities, train researchers, and even set up educational programs for high school students in California. The funding, $3 billion over a 10-year period, was considered enormous. California was becoming a mini-NIH for stem cell research. Moreover, because this proposition authorized the sale of general obligation bonds to fund the program, it was not subject to annual state appropriations. The new agency, California Institute for Regenerative Medicine (CIRM) would not have to compete for revenue with other state programs, especially during volatile economic downturns. This question of fiscal oversight, or lack of sufficient oversight, led to two lawsuits that were filed within months of Prop. 71's passage.[97]

One lawsuit was brought about by two anti-tax groups, People's Advocate and National Tax Limitation Foundation, both claiming that the new law was unconstitutional because it created a taxpayer-funded body, CIRM, that was not subject to state oversight or regulation. They also argued that the law set up an inherent conflict of interest because CIRM's grant review committee would include individuals who would be applying for grants themselves. The main issue, according to a lawyer for the plaintiffs, was that the "board that oversees the program is made up largely of university officials, patient advocates and biotechnology executives."[98] The plaintiffs argued that having university officials on the stem cell board was a conflict of interest because their institutions would be awarded many of the grants. The defense pointed out that the board members were appointed by state-elected officials and that the institute was subject to state financial audit procedures. The conflict of issues claim continued to haunt CIRM long after the lawsuit ended.

A second lawsuit, filed by the California Bioethics Council, was primarily concerned with the moral issues. The Council was affiliated with Focus on the Family, a pro-life organization that opposed hESC research because it destroyed human embryos. What was expected to be a long-drawn-out trial turned out to be shorter and less contentious. Prop. 71 was ruled constitutional. The decision was upheld on appeal, and when the state's Supreme Court refused to take the case, the California Institute for Regenerative Medicine (CIRM) was legally in business.

Actually, CIRM had been operating for several years using ad hoc funding from private donors as well as a loan from the state. It did not escape notice that Governor Schwarzenegger's decision to lend CIRM $150 million in July 2006 came one day after President Bush vetoed the Stem Cell Research Enhancement Act of 2005.[99] The governor had already signaled his support when he endorsed Prop. 71 in 2004.

CIRM's mission is to accelerate stem cell treatments to patients with unmet medical needs. When taking action, CIRM emphasizes that it considers whether that action will speed up the development of stem cell treatments, whether the treatment will be successful, whether it fulfills an unmet medical need, and whether it is efficient. How much money has been spent and who determines how the funds are allocated? Before we get to how the money was spent, it is important to look at CIRM's organizational structure. The conflict of interest charge brought about in the lawsuit discussed previously may become more evident.

The agency is governed by a 29-member Independent Citizens' Oversight Committee (ICOC) whose members are appointed for fixed terms by elected officials: the governor, the lieutenant governor, the treasurer, the controller, the speaker of the state assembly, and the president pro tempore of the state Senate. The idea here is that the oversight board is accountable to elected officials. Under Prop. 71, the governor, lieutenant governor, treasurer, and controller are responsible for nominating candidates for chair and vice-chair of the governing board. The ICOC then elects a chair and a vice-chair from among those nominees and selects a president who is the agency's chief operating officer. CIRM's first chair was Robert Klein, who served until 2011 and was replaced by Jonathan Thomas.[100] The appointment process, if a bit confusing, was created as such to promote transparency and accountability. But in fact, the law seemed to promote a conflict of interest by requiring that 23 of the 29 ICOC members had to come from "California institutions eligible for CIRM grants or of disease advocacy groups with their own interests in steering money toward their particular concerns."[101] The law makes it very clear that the money is to be spent for research being carried out by scientists working at universities and institutes located in California. Scientists could collaborate with counterparts in other states or even other countries, but the expectation was that results would accrue to the citizens of California. Every grant proposal submitted to CIRM requires each applicant to indicate how the citizens of California will benefit from his or her work. Despite an effort to promote transparency, CIRM was criticized for "embarrassing hints of cronyism," whether real or perceived.[102] CIRM adopted procedures to mitigate the appearance of impropriety, but criticism of the grant award process continued to trail the agency.[103] No one reading this book should be surprised to learn that Stanford University, a stem cell powerhouse, received a majority of CIRM grants, followed by labs located at University of California institutions. Ten universities make up the University of California system, and five of these have medical centers with extensive research facilities.

After almost 2 years in legal limbo, CIRM began funding grants for stem cell research in 2007. Money expended in 2006 was to secure research space. Like the NIH, CIRM also maintains a user-friendly database that can be searched by institution, researcher, grant type, and award value. The search can be refined through filters for disease focus (diabetes, Alzheimer's, etc.), stem cell use (hESC, iPSC, etc.), cell line generated, and collaborative funder.[104] CIRM encouraged joint funding for large projects such as construction of new facilities; a number of new centers were built to house research labs.

Prop. 71 provided CIRM with $3 billion over a 10 year period. The agency's funding is expected to continue through 2017, although their strategic plan extends through 2020. Table 5.4 provides an overview of total CIRM funding from 2006 to 2016. The agency did not award research grants until after their legal issues were resolved in 2007. In this table we also include NIH funding but only for hESC research. CIRM was established, in large part, to support hESC research when it was restricted by the Bush Administration.

Table 5.4 CIRM funding 2006–2016 (in millions of dollars)

	2006[a]	*2007*	*2008*[b]	*2009*[b]	*2010*	*2011*	*2012*	*2013*	*2014*	*2015*	*2016*
CIRM Funding	3	260	375	433	159	74	330	163	128	121	87
#Grants		156	97	75	51	43	54	36	37	19	15
NIH funding		42	88	143	166	123	146	146	166	180	191

Source: CIRM Annual Reports and database.

Notes

Total amount awarded between 2006 and 2016: $1.925 billion for 673 research grants.

a Operating funds from loans; no grants.

b Includes $271 million for construction.

Table 5.4 indicates that CIRM funded 156 grants in 2007 for a total of $260 million. Many of these early grants were educational—CIRM awarded predoctoral, postdoctoral, and clinical fellowships. It also awarded seed grants and creativity grants, which were typically in the $300,000 to $500,000 range, designed to help researchers jump-start their work. CIRM funded comprehensive research proposals comparable to the research project grants funded by the NIH. CIRM provides multi-year funding, typically for 3 to 5 years. In most years, CIRM allocated well over $100 million for research. Between 2008 and 2009, CIRM spent $271 million to help fund CIRM institutes at 12 universities or research centers. CIRM funding was only a small part of the total cost. The bulk of the money was provided by wealthy donors. For example, the Sanford Consortium in San Diego had been called the San Diego Consortium for Regenerative Medicine and located in a much smaller facility. CIRM contributed $43 million of the $163 million cost and T. Denny Sanford, a wealthy South Dakota businessman, put up $80 million. You might be able to figure out why the name changed. The Eli and Edythe Broad Foundation donated $75 million to help the University of California at San Francisco (UCSF) build the Broad Center for Regenerative Medicine and Stem Cell Research, with an additional $35 million from CIRM. Stanford University built the Lorry I. Lokey Stem Cell Research Building with a $75 million donation from Lorry Lokey, $44 million from CIRM, and the rest from university capital funds. The new agency did not seem to have much trouble convincing donors that their mission was a valuable one with great potential to advance medical science. CIRM was also criticized for putting up shiny buildings but not producing any stem cell therapies as promised.[105]

Table 5.4 also lists NIH funding for hESC during the same time period. CIRM was outspending NIH between 2007 and 2009, a period when the NIH was still operating under President Bush's funding limitation. The agency was just starting to award grant funding and, given the two-year hiatus due to legal battles, it had a backlog of applications. Although CIRM awarded a lot of small grants during this time period, it also supported a number of larger awards—between $1 and $3 million. CIRM was eager to fund translational research—either preclinical or clinical—and had an upper limit for proposals in these categories of $50 million.

CIRM's Centers for Excellence had project limits of $25 million, and CIRM special projects could be funded up to $10 million. These were, by any standards, pretty substantial figures. Our review of CIRM grants for the entire period under review (2007–2016) did not reveal any single grant in the $40–50 million range. However, a few multiple awards totaled about $40 million. In some cases, it was clear that the recipient had funds from other sources. For example, Pete Coffey, director of the London Project to Cure Blindness, who is running a clinical trial using hESC to cure blindness, was awarded two grants from CIRM (totaling $26 million), but he is also being funded by Pfizer, the New York Stem Cell Foundation, and funding groups in England. This is exactly the kind of research that CIRM wants to support. Once research makes it to clinical trials, the likelihood of becoming a viable treatment increases.

What is clear from the table is that CIRM funding is running out and that NIH's is increasing. With an annual budget of more than $30 billion, NIH will remain a major source of biomedical funding.

A second issue examined is the type of stem cells used in CIRM-funded research. A major reason for CIRM's existence was that funding for hESC research was limited. More research had been done using adult stem cells (especially blood stem cells), and research using adult cells was eligible for NIH funding. In fact, President Bush encouraged researchers to work with these cells.[106] Table 5.5 reports CIRM research funding by stem cell used.

Table 5.5 lists the number of grants and total funding by four stem cell types—hESC, iPSC, adult, and cancer. As expected, CIRM awarded more grants and funding for hESC research than research using any other type of cell. Many of these grants, however, were for training researchers as well as hiring new faculty. In absolute terms, CIRM did support hESC research. Grants for research using iPSC were second with 150 awards given out, but far less money was awarded ($330 million). There was great hope for iPSC research, but these cells

Table 5.5 CIRM funding and grant awards by types of stem cells used, 2007–2016 (funding in millions of dollars)

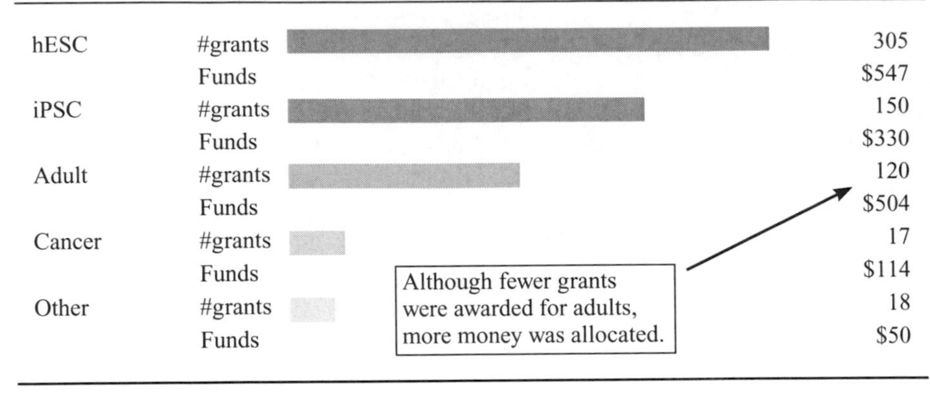

hESC	#grants	305
	Funds	$547
iPSC	#grants	150
	Funds	$330
Adult	#grants	120
	Funds	$504
Cancer	#grants	17
	Funds	$114
Other	#grants	18
	Funds	$50

Source: CIRM Grants database.

were only derived in 2006; scientists were still doing basic research. Finally, while only 120 grants used adult stem cells, the total monetary value was more than half a billion dollars. Our review of each of the 120 grants revealed that many of these grants used blood stem cells and were awarded for research that was in preclinical trials. Exploratory and training grants did not receive as much funding as research that was ready to be clinically tested.

The only way to get research out of the lab is to move to clinical trials. As we reported in Chapter 2, there are very few clinical trials using hESC; most clinical trials listed on the NIH registry are using adult stem cells.[107] Table 5.6 lists the 16 clinical trials funded by CIRM. Of the 16 trials, only 5 (4 are still active) are using hESC. Most of the trials are using adult stem cells. Two disease categories with more clinical trials, cancer and HIV/AIDS, are both using adult stem cells. The HIV/AIDS trials are especially significant since CIRM awarded more than $71 million to City of Hope, a nonprofit hospital and research facility in California that is running one of the trials. After reading several of the grant proposals from scientists at City of Hope, the importance of the HIV/AIDS trials are evident—14% of all the AIDS cases in the United States are in California. Only New York has a higher percentage. Finding a cure for this costly condition is a high priority for the state.

CIRM is again contributing to the funding stream for a new trial using hESC for spinal cord injuries. Asterias Biotherapeutics received a $14 million grant to begin the trial. Another

Table 5.6 CIRM clinical trials by stem cell type and disease and funding active and inactive trials

Adult Stem Cells Total		11	$millions
Immune	1		7
Heart disease	1		20
HIV/AIDS	3		31
Huntington's	1		9
Cancer	3		30
Leukemia	2		12
Embryonic Stem Cells Total		5	
Spinal cord	2		20
Diabetes	1		10
Blindness/AMD	2		18

Source: CIRM Grants database.

Note
Amount reflects CIRM funding only.

hESC trial that is getting a lot of attention: ViaCyte's, a California-based company, testing on diabetes. The company received a CIRM grant for $10 million for the clinical testing, but it received a total of $45 million over several grant cycles to conduct preliminary research. The current number of CIRM-funded clinical trials, while a bit skimpy, is moving the agency in an important direction if it is to achieve the objectives set out in its 2016–2020 strategic plan.[108]

The future of stem cell research: CIRM and NIH strategic plans

Both organizations issued 2016–2020 strategic plans, setting out their future goals and priorities. Although CIRM is primarily focused on stem cell research and NIH is not, both plans share some common features and concerns. Both emphasize the need to reduce the time-to-trial time frame, to push the translation of drugs and treatments, and to establish public–private partnerships.

The CIRM strategic plan, entitled *Beyond CIRM 2.0*, underscores the need to move beyond the lab and the clinic and get stem cell treatments into the market. The words "translation" and "accelerate" are used frequently throughout the plan. CIRM's founding mission is accelerating stem cell treatments to patients with unmet medical needs. After more than 10 years of awarding grants, there are still patients with unmet needs and treatments are few. CIRM's emphasis will be on developing stronger partnerships with the private industries that they believe can accelerate translation. To this end, CIRM announced in April 2016 that it had approved financing for the creation of a public–private company whose goal would be to develop these partnerships.[109] CIRM has awarded grants to a number of biotechnology companies, but, with few exceptions, the funding has not led to any clinical trials and, as yet, no commercial treatments. This result may take longer than the 5 years covered in the plan. California is home to hundreds of biotechnology companies, but many depend on some public philanthropic funding. Geron, a private company that initiated the first hESC trial in 2009, was funded by CIRM, as are two companies with current hESC clinical trials.

A second theme expressed in the *CIRM 2.0* plan is the need to change the regulations governing clinical trials and approval for drugs and treatments. The main antagonist here is the U.S. Food and Drug Administration (FDA). CIRM and NIH both refer to the "valley of death" as the time it takes from drug/treatment discovery to final approval.[110] CIRM strategy is to work with the FDA to find ways to accelerate the approval process for clinical trials and for final approval of new drugs and therapies. There are pluses and minuses of speeding up the drug/treatment approval process.

The NIH's strategic plan, entitled *Turning Discovery into Health*, includes a much larger research mission.[111] In his introduction to the plan, the NIH director notes that "working with our many partners in the public and private sectors, NIH will use this framework as we strive to turn scientific discovery into better health…."[112] The NIH is also concerned that it takes too long to move from discovery to translation to clinic. The NIH provides a translation timeline underscoring that the development of a new therapeutic is long (about 14 years) and costly (about $14 billion). Reducing these figures is something that NIH is promoting through their Clinical and Translational Science Awards program; an alternative way to design clinical trials that permits flexibility but still ensures patient safety.

The NIH funds research conducted by scientists working at private companies. Like CIRM, NIH wants to encourage more public–private partnerships with health-related companies and small businesses. In 2014, NIH established the Accelerating Medicines Partnership, which includes the FDA, 10 biopharmaceutical companies, and a number of nonprofit organizations to find ways to increase discoveries and cut time to trial. The 2020 plan proposes to push this partnership to increase results.

Both strategic plans emphasize speeding up the discovery to result timeline. The push in California is more vital since CIRM's funding will run out (it was funded for 10 years). The agency would like to be credited with a few commercial treatments. It remains to be seen if any of the CIRM clinical trials will produce the expected breakthrough. The NIH is not likely to run out of money, but as its strategic plan points out: "NIH funding has not kept pace with inflation, and the agency has lost approximately 22% of its research purchasing power since 2003."[113] It may take longer for some of the expected breakthroughs to emerge as fast as the NIH plan projects.

Although there is great hope that stem cell therapies in general, and hESC in particular, will emerge in the next 5 years, given the limited number of clinical trials, it may take another decade. The issues of time and cost for stem cell research are continuous. There are always more scientists asking for money than there is money to disburse. A related issue now facing CIRM is what happens when the money runs out. In June 2016, at a meeting of the International Society for Stem Cell Research (ISSCR), one of the topics discussed was the hype surrounding stem cell research that had led to the proliferation of unapproved stem cell clinics around the world. In an earlier chapter we discussed stem cell tourism. But the headline that caught our attention read: "Another way to dial back stem cell hype (but not hope): Put a dollar figure on it."[114] The blog, published on CIRM's homepage, reported on a presentation by a British stem cell researcher, Roger Barker. Barker, who works on Parkinson's disease, proceeded to discuss the various costs associated with research going from lab to clinical trial. Even for a phase I trial with 20 patients, he estimated costs between $3–5 million for the basic research, plus another $3 billion to go from phase I (safety) to phase III (efficacy and large trial). He ended his presentation by adding, that even if the study proves to be successful, you still have to prove that the new therapy is more effective and cheaper than the existing therapy. The bottom line is that California's $3 billion investment in CIRM is barely enough to bring a few therapies to market.

At this point you might be wondering if it is so costly to conduct stem cell research, why are states and the federal government spending so much taxpayer money on this effort? An easy answer is to borrow part of the NIH's mission statement and say that money is spent "to enhance health, lengthen life, and reduce illness and disability."[115] Earlier in the chapter we cited some statistics that confirm that we live longer and suffer less due to scientific discoveries that have been translated into drugs and therapies. But we might want to explore another line of thinking that argues for privatizing science research.

Private companies (especially multinational pharmaceutical companies) have budgets larger than many state governments; private companies know how to make and market products and are already involved in stem cell research. Several articles that appeared during the Bush Administration years when hESC funding was limited argued that "because stem cell research is inherently speculative and politically controversial, the public would be best served if governments left it to the private sector."[116] Others have confirmed that private companies and even wealthy donors have stepped in to fund hESC research.[117] Between 2005 and 2007, six private companies that were engaged in stem cell research increased their research and development expenditures from $64 million to $156 million.[118] During this same time period, private donors and disease-specific foundations (Juvenile Diabetes Research Foundation) contributed more than $1 billion for research.[119] While these expenditures are impressive, it is difficult to anticipate that future contributions will be sustained beyond the first year. CIRM did not seem to have much trouble obtaining private funding for the construction and naming of research facilities. Operating a research lab, however, requires an annual revenue source. The case for letting the private sector fund hESC research is based, in part, on the general affirmation that the private sector will do it and that the private sector is more efficient (it can do more with less money).

A second part of the claim for reducing government spending and regulation is based on the example of IVF technology, which the federal government declined to fund in the 1980s. The IVF industry is now a multibillion dollar enterprise with minimal federal regulation. In 1992, the federal government mandated that IVF clinics report their success rates to the CDC to reduce misleading claims.[120] While critics of the procedure continue to cite moral concerns, there is no movement to close the clinics or to exert more regulatory control.[121] Some scholars point out that stem cell science (without federal funding) might have gone the way of IVF.[122]

The private sector is quite involved in hESC research. We know that Geron funded research on the derivation of the first hESC lines as well as the first clinical trial. Private donors helped pass Prop. 71 in California then paid for the agency to start operating while pending the outcome of a lawsuit. Private donors also contributed heavily to the construction of many new stem cell research centers in California as well as other states. We also read about the search by both CIRM and NIH for private sector partners to take the research to the next step. So far there have been no attempts to keep the private sector out of this arena. It is difficult to say, at this stage, what would have happened if President Bush had decided to prohibit any hESC research. Would we have private stem cell clinics advertising treatments for various ailments? Yes, and, we do. One of the unintended consequences of the hype over hESC research is the proliferation of stem cell clinics across the United States that are marketing unproved treatments.

In Chapter 2 we discussed stem cell tourism as something taking place in other countries.[123] But a recent article published in the journal *Cell Stem Cell* confirms that there are, "351 U.S. businesses engaged in direct-to-consumer marketing of stem cell interventions offered at 570 clinics."[124] And where are most of these unregulated clinics? You guessed it—California, more precisely Beverly Hills. For the most part these clinics are offering commonly used cosmetic procedures that use the patient's own fat cells, but some are also performing bone marrow treatments. This finding raises the question of what to do about these clinics? Newspaper articles report that the FDA is scheduling public hearings to review the data and get public comments. One side says regulate. The other side says do not regulate. These clinics are simply engaged in the "practice of medicine."[125] Plastic surgeons regularly inject patients with substances like Botox as well as the patient's own fat to help them regain a youthful appearance. If it is these procedures that clinics are promoting using a different "frame," then this might not be a problem. One might even be persuaded that if the unproved treatments do no harm, then they should be allowed to continue. Or you might be concerned that some clinics are engaged in deceptive advertising and should be regulated, or at least required to include a cautionary note or disclaimer in their glossy magazine features. The FDA is "concerned that the hope patients have for treatments not yet proven to be safe and effective may leave them vulnerable to unscrupulous providers of stem cell treatments that are illegal and potentially harmful."[126] Is the FDA really worried about quashing the "hope" of patients, or is the real concern that an unproved and unregulated therapy will cause irreversible damage or even lead to the death of a patient? And, if and when that happens, who will/should be held responsible? Who will/should be sued?

We do not have an answer for this emerging issue. The discovery of stem cell clinics leads some members of Congress to contemplate "allowing the FDA to approve stem cell treatments for five years without advanced trials."[127] That, it seems, is exactly what CIRM and the NIH are asking for in their strategic plans. Revising the clinical trial phases (see Chapter 2) would free up a lot of money now spent conducting lengthy and elaborate clinical trials. Can we interest you in a stem cell treatment? No guarantees that you will get better, but hopefully you will not get worse.

At this point, no one is asking the federal government or the states already funding their own stem cell research efforts to turn it all over to the private sector. What we do see is that very partnership that has always existed between agencies engaged in the research of discovery and private companies willing to move some discoveries to the market. One of the risks of relying solely on the private sector is that the research may focus on diseases more likely to produce a profit. Diseases or conditions that affect a small number of individuals may be neglected. Several of the hESC clinical trials are targeting age-related macular degeneration (AMD). Given the increase in older individuals in the United States and other countries, the demand for this treatment is likely to be high. The migration of stem cell research into the medical market will probably be financed by private companies and/or investment groups.

In the final part of this chapter we revisit and expand upon some of the nonmonetary consequences produced by hESC research. Perception is often as important as reality. This was confirmed recently in the previously cited study about the proliferation of stem cell clinics in the United States. These clinics are a manifestation of the "hope patients have for treatments not yet proven."[128] We might go a step further and say hope for treatments that have yet to be developed. The discovery of hESC in 1998 was followed by years of media reporting that without federal funding the miracle cures would not be developed. Or they would be developed but by scientists in other countries. Now almost 20 years later we are still waiting for the miracle cures. In Chapter 4, we confirmed that a majority of the public supports public funding of hESC. The consequence of overselling the science created rising public/patient expectations. Recognizing that the explosion of news coverage has contributed to the hype, the International Society for Stem Cell Research (ISSCR) reissued its publication, *Guideline for Stem Cell Research and Clinical Translation*, in May 2016.[129] The revised publication includes a new chapter on communication, in which the ISSCR "explicitly recognizes and confronts the issue of science hype."[130] The guidelines endorse the following[131]:

> Recommendation 4.1: The stem cell research community should promote accurate, balanced, and responsive public representations of stem cell research.
> Recommendation 4.2: When describing clinical trials in the media or in medical communications, investigators, sponsors, and institutions should provide balance and not emphasize statistically significant secondary results when pre-specified primary efficacy results are not statistically significant. They should also emphasize that research is primarily aimed at generating systematic knowledge on safety and efficacy, not therapeutic care.

The research community is encouraged to realistically portray their work in a way that does not contribute to the public's unrealistic expectations. In translating the recommendations into action items, one writer suggests that researchers should reign in sensational language, limit glowing single-patient anecdotes, include limitations, risks, costs, and harms, and ease up on the problems of stem cell tourism and pseudomedicine clinics.[132]

However, many journalists, even science journalists, cannot resist including a story about a "family's or person's desperate hope for a cure."[133] A good story with a happy ending is preferable to one that reveals the negative side effects or that the patient saw no change despite spending a lot of money.

Brain drain and policy uncertainty

The tendency for researchers to oversell the potential results is one, and maybe the most, damaging, nonmonetary consequence of stem cell research. The two other consequences mentioned by President Obama are "some of our best scientists leave for other countries…" and, "those countries may surge ahead of ours."[134] The brain drain out of the country never materialized, although some scientists did leave and funding issues appear to have created uncertainty. But there was no mass exodus. Stem cell research is a global field and collaboration across national borders is common.

A dictionary definition of brain drain is "the situation in which large numbers of educated and very skilled people leave their own country to live and work in another one where pay and conditions are better."[135] This definition perfectly describes stem cell researchers. These individuals are very educated and skilled. And while pay is an important reason to change jobs, more important for these researchers is the opportunity to work with hESC. But initially there were not that many hESC scientists. A few labs around the world had been working simultaneously at deriving stem cell lines when the first lines were isolated in the United States. James Thomson and John Gearhart were the founding biologists who ushered in a new field of science. It was expected that the United States would retain the secret of how to develop these "miracle cells" into useful medical treatments. This is not how science works. Scientists are eager to share their findings and accomplishments. There were other scientists in other countries working on the same research. Replication is a foundation of science. Among the first hESC lines approved by President Bush in 2001 were lines from India, Sweden, South Korea, Israel, and Australia in addition to the United States. We would not expect that these countries would put their research on hold while the United States moved ahead on developing medical applications. The International Society for Stem Cell Research was founded in 2002 for the express purpose of promoting "global collaboration among the world's most talented and committed stem cell scientists and physicians."[136]

Although President Clinton seemed to support the research, the existence of the Dickey–Wicker Amendment cast a shadow over the research, as evidenced by the fact that both Thomson and Gearhart used private funds to create their hESC lines. Even after President Bush decided on limited NIH funding, many researchers continued to rely on private funding to establish their labs. However, the word "uncertainty" is one that is frequently associated with stem cell research in those early years and even into the Obama Administration. An article entitled "Policy uncertainty and the conduct of stem cell research" sums up the cloud that followed some scientists.[137]

First, there was the uncertainty of if NIH would fund any research under the 2000 guidelines. NIH did not. Second, there was the uncertainty of if President Bush would approve federal funding for any hESC research. He did (but limited lines that could be used). Third, there was the uncertainty of if the Supreme Court would uphold the legality of President Obama's executive order issued in 2009. After 2 years in limbo, the Court refused to hear the case. And, in California, there was the uncertainty of whether or not the state Supreme Court would uphold the legality of Prop. 71. After almost 2 years, the state Supreme Court refused to hear the case. If you were a stem cell scientist, how stable and comfortable would you feel doing research in this volatile environment? We can only guess that it would be somewhat disconcerting. Most scientists, even those with university teaching positions, are dependent on external funding to support their labs and their assistants. A tenured faculty position does not usually come with open-ended research funding. Universities might use private donations or their capital budgets to fund the construction of a new building or lab, but it is the principal investigator who is responsible for determining the direction of the lab's research and writing grant proposals to support the lab and all the employees. Seeing a scientific article with a dozen or more coauthors confirms that research is not a solo effort. The effort is expensive.

How many researchers left the United States in the wake of an impending ban on federal funding is hard to determine with any degree of quantitative accuracy. Like other professionals, scientists move for a variety of reasons. One well known scientist, Roger Pedersen, did leave and became a symbol for the expected brain drain due to the U.S. policy. It is hard to find a book[138] or article[139] on stem cell research that does not mention his name and why he left the United States for England. Like Thomson and Gearhart, Pedersen had derived his own hESC lines with money from Geron—it just took him a little longer. Pedersen applied for a NIH grant in 2000, during that small funding window between the Clinton and Bush Administrations. No funds were ever awarded. Pedersen, who had been on sabbatical in 2000, recalled that "shortly after I got back to UCSF from Cambridge in early 2001, I had a phone call from NIH headquarters to say they were terminating my application. I later learned that this reflected NIH's imposition of the newly elected Bush Administration policies for stem cell funding. That's when I began thinking seriously about moving to England."[140]

Pedersen resigned from his position as director of reproductive genetics at UCSF, making it clear he was leaving for "the possibility of carrying out my research with human embryonic stem cells with public support."[141] Pedersen became a symbol for the unsettled state of U.S. policy on stem cell research. According to the dean of the UCSF medical school, "if federal support for stem cell research is not forthcoming, the risk exists that talented scientists will leave academic centers to seek opportunities in the private sector, or even overseas."[142] In an interesting twist of fate, in 2009 when President Obama expanded funding for hESC, the British Royal Chemical Society bemoaned in a newsletter that the funding boost raised concerns about a brain drain from Europe to the United States. So real or perceived, the idea of scientists leaving is a global concern.[143]

Two other well known scientists, Neal Copeland and Nancy Jenkins, left their jobs at NIH's National Cancer Institute in 2005 to set up their lab at Singapore's Institute of Molecular and Cell Biology. The husband-and-wife team of cancer geneticists had been offered jobs at Stanford University but decided that the uncertainty over the implementation of Prop. 71 was unsettling.[144] In 2011, they both returned to the United States, commenting that they could no longer do their work in Singapore because the city-state was shifting its funding priorities to research with industrial applications.[145] Both are now working at the newly created Methodist Hospital Research Institute in Houston, Texas. This may be evidence of a reverse brain drain back to the United States. Or, it may reflect the ebb and flow of the recruitment and retention patterns of well known experts in any field who are presented with attractive job offers. There are many examples of stem cell researchers with international reputations who were recruited by U.S. universities or agencies. Alan Trounson, the first president and CEO of CIRM, was the founder of the Australian Stem Cell Centre in 2003; in 2004, he established the Monash Immunology and Stem Cell Laboratories. He was a very experienced and established researcher before he left for California.[146] Pick a stem cell biology or regenerative medicine program at any major U.S. university and read the resumes of the faculty. You will find an international collection of scientists who came to the United States to pursue their passion.

Some scientists may have threatened to leave, giving rise to the rumor of an even larger exodus than actually occurred. The example of Jerry Yang, an animal cloning expert at the University of Connecticut, is illustrative of the willingness of one state to retain their international star. Yang, who was Chinese, had been working collaboratively with universities and research centers in both China and Japan cloning farm animals. Rather than risk losing him, the university spent $10.6 million to open the Advanced Technology Laboratory, which gave Yang and his lab members an opportunity to push stem cell technology to the next level.[147] Yang passed away in 2009 just as the political climate was changing, which would have allowed him to work on human embryos in addition to the animal embryos that were his specialty.[148]

The United States was not the only country that feared a brain drain in the early years of hESC research. Europe experienced a similar situation. When the European Council of Science Ministers issued a one year moratorium on hESC funding in July 2002, the fear was that "stem cell scientists would leave for the United Kingdom, Australia, and Singapore, where such work is legal, and also to the U.S., where work can be done with private funds."[149] Christine Mummery, a well known Dutch stem cell scientist, is reported to have exclaimed upon hearing the news: "It's an absolute disaster.... There are certain groups within Europe—Germany and France—asking whether people can come to our lab and work because they don't want to get behind with expertise."[150] In addition to uncertainty about funding, Mummery confirms the need for scientists to maintain their skills and keep up with new research. Faced with the possibility of an absolute ban on all hESC research, even the United States looked like a desirable option. In fact, the European Union did not extend the ban, rather it permitted individual countries to adopt different regulations regarding hESC research. For example, the derivation of embryonic stem cell lines is not permitted in Germany and Italy, but researchers can import hESC lines from other countries. In 2004, the European Consortium for Stem Cell Research (EuroStemCell) was created and funded by the European Commission.[151] This is a partnership of more than 400 stem cell research and regenerative medicine labs across Europe. It is largely a clearinghouse for information about stem cell research, clinical trials, and funding opportunities. It maintains a very up-to-date webpage with information provided in a number of languages.

In an effort to understand the effect of policy uncertainty on stem cell researchers, one U.S. study reported on the results of a survey sent to scientists. Although the sample included scientists who were working with adult stem cells as well as those working with hESC, the findings confirm the hypothesis that uncertainty extended into the Obama Administration when a legal challenge temporarily prevented the disbursement of funding.[152] The most common impact of the uncertainty was a delay in planning future research and an obstacle to completion of ongoing work.

The mass exodus of stem cell scientists from the United States cannot be confirmed, but the uncertainty created by ever-changing policies both at the federal level and state level casts a pall over the field. This pall was not permanent, as evidenced by the increase in the number of scientists applying for NIH grants (see Table 5.4). In addition, the first PhD program in stem cell biology and regenerative medicine was created in 2011 at Stanford University. The new program was designed to "accelerate the translation of discoveries into clinical therapies."[153] It is increasingly evident (and maybe necessary) that scientists associated with stem cell research were voicing an interest in translation.

National prestige and stem cell funding

We conclude this chapter by revisiting the issue of national prestige and science funding. The chapter focused on NIH funding for hESC and the effort by states, especially California, to support their own program. To what extent was national prestige diminished because of limits on federal funding? National prestige is an elusive concept. It all depends on who you ask and when. It also depends on context and content. Biotechnology is not something that the average citizen may know a lot about, but when asked if they believe their country is a leader in the field, most are likely to reply in the affirmative.

In 2005, Research!America, a large, nonprofit organization dedicated to making research on improving health a higher national priority, included a question on one of their national surveys. The question: How important do you think it is that the United States is a global leader in medical and health research? As you might expect, 78% of the respondents said very important. That exact question was not repeated in subsequent years. However, in both 2012 and 2016, a question about country status was included. The question: In your view, which of the following (countries) will be considered the number one leader in science and technology in the year 2020? In 2012, 41% of the respondents said the United States, followed by 33% who said China, and 30% who chose the "not sure" option. In 2016, 44% thought the United States would be the leader, 19% said China, and 22% were not sure.[154] Other countries, such as India, Brazil, and the European Union were included, but responses were low. The importance of being a global leader is clear; not so clear is that the United States will be able to achieve that status by 2020. The other explanation is that the average citizen does not know enough about medical breakthrough to say which country will be ahead. A third explanation might be that medical technology has become a global commodity; it will be available irrespective of which country discovers the drug or procedure. Scientists working in the field were equally conflicted about the status and prestige of the United States as a leader in stem cell research.

In 1999, a year after Thomson and Gearhart announced that they had isolated the first hESC lines, the American Association for the Advancement of Science proclaimed the discovery as the "breakthrough of the year."[155] U.S. scientists had accomplished what their counterparts in other countries had yet to achieve. The prestige of America and its scientific community was high. In 2001, *Time* magazine featured a picture of James Thomson with the byline: "the man who brought you stem cells is one of America's best in science and medicine."[156] This would seem to be a clear endorsement that America possessed a scientific community that was the envy of the world. The prestige of the United States was riding on the expectation that the science of discovery that these men and women were working on would soon deliver clinical cures. In some respects, this early period could be compared to President John F. Kennedy's declaration in 1961 that the United States should commit itself to landing a man on the moon. The president asked for billions of dollars to achieve this goal. In 1969, the United States did land a man on the moon.

But the United States was not committed to spending billions of dollars on stem cell research, in part due to the moral conundrum expressed by President Bush and shared by some Americans. Between 2001 and 2005, U.S. scientists began to question whether the United States was abdicating its position as the world's preeminent biotechnology leader to other countries. Scientists who left the United States for other countries, like Roger Pedersen, warned that the United States may be left behind when it comes to future medical advances. Others offered a bleak scenario that "the United States has steadily fallen behind several other countries, such as Britain, Israel, and Singapore, that are rapidly moving ahead in this field."[157] South Korea could also have been added to this list of countries where the stem cell breakthroughs would likely occur. At the first meeting of the International Society for Stem Cell Research in 2003, the South Koreans presented 10 poster sessions on hESC research; the United States presented 7 and the United Kingdom none.[158] A year later, South Korea would make history by claiming (falsely it was later revealed) that they had successfully created stem cell lines from cloned embryos, boosting the prospect that patient-specific therapies were around the corner.[159]

Newspapers around the country proclaimed that the United States was falling behind other countries. An editorial in *USA Today* announced that it was "time to put the U.S. at the forefront of promising research," stating that the "reasoning behind Bush's limits on stem cells no longer makes sense."[160] A statement from the Coalition for the Advancement of Medical Research noted that "we have spent the last 50 years in this country building a biomedical research enterprise that's the envy of the world, but with stem-cell research, we are giving that lead away."[161]

The South Korean cloning announcement was a major accomplishment, a real cloning coup. U.S. scientists acknowledged that "it's unbelievable good news for patients," and that the work was a "major medical milestone."[162] However, you could almost feel the sting of resentment as some ponder whether the prestige of the United States as a biotechnology powerhouse would diminish. They lamented that a "technology largely created in the richest nation on earth was getting more support abroad."[163] Stephen Minger, an American stem cell researcher who was working in England, observed that "Hwang's laboratory is now way ahead of the field.... There is a good chance that the U.S. will be left behind as the situation on stem cell research there becomes more fragmented and incoherent."[164] In 2009, Minger returned to the United States to work for an American company.

In the United States, the federal government did not spend millions of dollars to construct stem cell hubs as South Korea and Singapore did. Some states, using a combination of public and private money, did set up new facilities dedicated to stem cell research. These centers in California, Massachusetts, Connecticut, and New Jersey are but a few of the efforts underway in the United States to move stem cell research to the next level. When it was revealed that the South Koreans had faked the data, U.S. scientists as well as some members of Congress reiterated the claim that the United States was losing its science edge and began to again lobby for more hESC funding. They got their wish in 2009. However, the world of stem cell research had changed by 2009. A Japanese scientist had figured out how to take adult stem cells, reprogram them using four specific genes, and turn the cells back to their embryonic-like state. Shinya Yamanaka, the Japanese orthopedic surgeon turned stem cell scientist, won the Nobel Prize in 2012.[165] Should the American, James Thomson, have won the Nobel Prize? That would certainly have reaffirmed the United States' position as a stem cell powerhouse. Both Yamanaka and Thomson shared the honor of being on *Time* magazine's 100 most influential people in 2008,[166] Yamanaka for having figured out how to derive iPSC initially from adult mouse cells and later human cells and Thomson for having figured out a different procedure for deriving human iPSC. Both scientists have established research connections in California (Thomson at University of California, Santa Barbara and Yamanaka at the Gladstone Institute at the

University of California, San Francisco). The global pursuit of stem cell science is breaking down national boundaries.

Summary

The demands for increased stem cell research funding are likely to continue into the future. The concern that other countries such as China, India, and Europe are investing more in this research were repeatedly voiced when NIH funding was limited.[167] Recently, evidence that private biomedical companies are investing more of their research and development dollars in Asia than the United States has renewed the call for boosting U.S. government funding.[168] The call for increased funding to advance stem cell science is likely to continue. The presentation of new proven stem cell therapies will increase the public's willingness to support budget increases. Until then, look for the debate to continue.

Additional readings

Original sources and other scholarly readings

Scholarly research requires reading original sources in addition to secondary sources. This may include going to the original funding figures provided by federal and state agencies.

1. FASEB, *Federal Funding for Biomedical and Related Life Sciences Research FY 2017*; http://www.faseb.org/Portals/2/PDFs/opa/2016/Federal_Funding_Report_FY2017_FullReport.pdf.
2. FASEB, *Federal Funding Data*; [interactive map] http://www.faseb.org/Science-Policy-and-Advocacy/Federal-Funding-Data/Federal-Funding-by-State-and-District.aspx.
3. National Conference of State Legislatures, *Embryonic and Fetal Research Laws*, January 1, 2016; http://www.ncsl.org/research/health/embryonic-and-fetal-research-laws.aspx.
4. Council of State Governments, *Book of the States, 2016*; http://www.csg.org/policy/publications/bookofthestates.aspx (This book is available at most university libraries. It provides comparable information for each of the states. A good source of original data.)
5. U.S. Department of Commerce, United States Census Bureau, *United States Census 2010*; http://www.census.gov/2010census/; also, http://www.census.gov/fastfacts/.
6. A. D. Levine, Policy uncertainty and the conduct of stem cell research, *Cell Stem Cell* 8, February 4, 2011: 132–135; http://www.ncbi.nlm.nih.gov/pubmed/21295270.
7. L. Turner and P. Knoepfler, Selling stem cells in the USA: Assessing the direct-to-consumer industry, *Cell Stem Cell* 18, June 30, 2016; http://www.cell.com/cell-stem-cell/fulltext/S1935-5909(16)30157-6.
8. A. D. Levine and L. E. Wolf, The roles and responsibilities of physicians in patients' decisions about unproven stem cell therapies, *Journal of Law, Medicine, & Ethics* 40(1), Spring 2012: 122–134; http://www.ncbi.nlm.nih.gov/pubmed/22458467.
9. D. Lau et al., Stem cell clinics online: The direct-to-consumer portrayal of stem cell medicine, *Cell Stem Cell* 3, December 4, 2008: 591–594; http://www.ncbi.nlm.nih.gov/pubmed/19041775.
10. D. F. Maron, Unproved stem cell clinics proliferate in the U.S., *Scientific American*, June 30, 2016; http://www.scientificamerican.com/article/unproved-stem-cell-clinics-proliferate-in-the-u-s/.
11. American Academy of Arts & Sciences, *Restoring the Foundation: The Vital Role of Research in Preserving the American Dream*, Cambridge, MA, 2014; www.amacad.org/content/Research/researchproject.aspx?d=1276.
12. International Society for Stem Cell Research, *Guideline for Stem Cell Research and Clinical Translation*, May 12, 2016; http://www.isscr.org/docs/default-source/guidelines/isscr-guidelines-for-stem-cell-research-and-clinical-translation.pdf?sfvrsn=2.

13. California Institute for Regenerative Medicine, *Beyond CIRM 2.0: Proposed Strategic Plan 2016 and Beyond*; https://www.cirm.ca.gov/sites/default/files/files/agenda/151217_Agenda_7_CIRM_StratPlan_final_120815.pdf.

Secondary analysis and news articles

1. L. McGinley, Unregulated stem-cell clinics proliferate across the U.S., *Washington Post* A3, July 1, 2016; https://www.washingtonpost.com/news/to-your-health/wp/2016/06/30/unregulated-stem-cell-clinics-are-proliferating-across-the-u-s/.
2. S. Len, South Korea, with renowned scientists, jolts field and revives debate, *New York Times* A2, February 13, 2004; http://www.nytimes.com/2004/02/13/us/cloning-stem-cells-laboratory-south-korea-with-renowned-scientists-jolts-field.html.
3. K. Kaplan, Hundreds of companies in the U.S. are selling unproven stem cell treatments, study says, *Los Angeles Times* June 30, 2016; http://www.latimes.com/science/sciencenow/la-sci-sn-unapproved-stem-cell-treatments-20160630-snap-story.html.
4. K. Boadi, Erosion of funding for the National Institutes of Health threatens U.S. leadership in biomedical research, *Center for American Progress*, March 25, 2014; https://www.americanprogress.org/issues/economy/report/2014/03/25/86369/erosion-of-funding-for-the-national-institutes-of-health-threatens-u-s-leadership-in-biomedical-research/.
5. *CBS News*, *Stem Cell Fraud*, January 12, 2012; https://www.youtube.com/watch?v=ovPZkQYee8Y.
6. Stem Cells in Translation, *Cell* 153, June 6, 2013: 1177–1179; http://www.cell.com/cell/issue?pii=S0092-8674(13)X0012-1.
7. S. Minger, *Blue Skies—The Future of Regenerative Medicine*, February 1, 2015; https://www.youtube.com/watch?v=w5Yohe2jZd4 [Stephen Minger, Chief Scientist, Life Sciences, GE Healthcare].

Critical thinking activities

1. After reading this chapter you are familiar with the different sources of funding for basic research. With an annual budget of more than $30 billion, NIH provides significant support for science research. The Federation of American Societies for Experimental Biology (FASEB) provides a user-friendly interactive website that summarizes federal science funding by state and district. Go to the webpage and practice using the interactive map at: http://www.faseb.org/Science-Policy-and-Advocacy/Federal-Funding-Data/Federal-Funding-by-State-and-District.aspx.

   Option 1. Pick a state that supported funding for hESC and one that did not or one that neither supported nor prohibited the research (see Table 5.3). The map includes data on total state funding, the number of institutions receiving funding, the total number of grants, and the total number of individual awards. You should plan on doing some additional research on each of the states, such as finding out total population, per-capita income, total number of universities and hospitals, etc. Census data will be a good starting point at: www.census.gov/fastfacts. Visit the webpage of the individual states you have selected. You might also find useful information at the National Conference of State Legislatures at: http://www.ncsl.org/.

   Compare the two states with respect to NIH funding. Develop and support a thesis based on your findings. What type of additional data would you like to have to further support your thesis?

   Option 2. Pick two states you are interested in learning more about. Look at the total distribution of federal research money received by the states. You should plan on doing some additional research on each of the states such as finding out total population, per-capita; income, total number of universities and hospitals, etc. Census data will be a good starting point at www.census.gov/fastfacts. Visit the webpage of the individual states you have selected. You might also find useful information at the National Conference of State Legislatures at: http://www.ncsl.org/.

Do you see any differences between the two states? In examining the data you may find that some agency other than NIH provides the state with substantial funding. What type of research does the funding support? Is agriculture a major source of income for the state? Develop and support a thesis based on your findings.

2. The proliferation of stem cell clinics was discussed in this chapter. It was once assumed that most of these unregulated clinics were in foreign countries, but a recent study confirms that many are in the United States. Begin by reading the original study published in the peer-reviewed journal *Cell Stem Cell*. Then read an article in *Scientific American* that reported on the study. Finally, read a newspaper article that summarized the findings. First, write an essay in which you compare the different ways in which the three readings relay information to their audience. How different, for example, are the titles of each reading? Second, try your hand at content analysis. Using the same search term used by Turner and Knoepfler, do your own search. Find and analyze 10 of the websites you find that offer stem cell treatments. Is it clear to you what the clinics are providing? Is the clinic simply renaming their product to include the term stem cell (a kind of re-framing that was discussed in Chapter 3)? Or is the clinic selling a procedure that cannot deliver the claimed result? Is the procedure a routine one (injection of Botox or fat cells), or is the procedure more invasive requiring hospitalization? You may want to begin by watching a short segment done by the TV news program *60 Minutes* in 2012. Write an essay discussing your findings.
   a. *CBS News*, *Stem Cell Fraud* (January 12, 2012); https://www.youtube.com/watch?v=ovPZkQYee8Y.
   b. L. Turner and P. Knoepfler, Selling stem cells in the USA: Assessing the direct-to-consumer industry, *Cell Stem Cell* 19, August 4, 2016; www.cell.com/cell-stem-cell/flltex/S1934-5909(16)30157-6.
   c. D. F. Maron, Unproved stem cell clinics proliferate in the U.S., June 30, 2016; http://www.scientificamerican.com/article/unproved-stem-cell-clinics-proliferate-in-the-u-s/.
   d. K. Kaplan, Hundreds of companies in the U.S. are selling unproven stem cell treatments, study says, *Los Angeles Times*, June 30, 2016; http://www.latimes.com/science/sciencenow/la-sci-sn-unapproved-stem-cell-treatments-20160630-snap-story.html.
3. Understanding the science in an hour from one of the top stem cell researchers in the United States. Listen to Dr. Douglas Melton, professor of Stem Cell and Regenerative Biology at Harvard University and codirector of the Harvard Stem Cell Institute, give a video lecture on *Understanding Embryonic Stem Cells*. After watching the lecture, pick two of the segments in which he discusses some of the issues that continue to puzzle scientists (the lecture is divided into about 40 substantive subsections). After doing some additional research, write a three-page paper in which you discuss why scientists are still trying to figure out some of the key issues that he brings up. Hint: differentiation of hESC. https://www.hhmi.org/biointeractive/understanding-embryonic-stem-cells.

Notes

1. B. Obama, Remarks on signing an executive order removing barriers to responsible scientific research involving human stem cells and a memorandum on scientific integrity. *Public Papers of the Presidents of the United States*, Book 01, March 9, 2009, 199.
2. Ibid.
3. Ibid.

4. National Institutes of Health, National Cancer Institute. *National Cancer Act of 1971*, S. 1828, December 23, 1971, Pub. L 92-218.
5. U.S. Department of Health and Human Services, Centers for Disease Control and Prevention, National Center for Health Statistics, Health United States, 2015, Table 19, p. 107.
6. L. McGinley, Biden unveils launch of major, open-access database to advance cancer research, *Washington Post*, June 6, 2016.
7. J. A. Thomson et al., Embryonic stem cell lines derived from human blastocysts, *Science* 282, November 1998: 1145–1147; and, M. J. Shamblott et al., Derivation of pluripotent stem cells from cultured human primordial germ cells, *Proceedings of the National Academy of Sciences* 95, November 1998: 13726–13731.
8. The strategic plans for both the California Institute for Regenerative Medicine (CIRM) and the National Institutes of Health (NIH) underscore translation as a key variable in funding decisions. CIRM, *Beyond CIRM 2.0: Proposed Strategic Plan 2016 & Beyond*; also, NIH, *NIH-Wide Strategic Plan: Fiscal Years 2016–2020*.
9. D. M. Herszenhorn and C. Hulse, Deal struck on $789 billion stimulus plan, *New York Times*, February 12, 2009.
10. D. Matthews, The sequester: Absolutely everything you could possibly need to know, in one FAQ, *Washington Post*, February 20, 2013.
11. The authors argue that government funding of stem cell research is "bureaucratic, wasteful, and mired in political controversy" (p. 1). They believe that private biotechnology companies along with philanthropic donations are more than adequate to fund hESC research. Between 2006 and 2007, the expenditures of six companies totaled more than $279 million. Philanthropists such as Michael Bloomberg donated $100 million to Johns Hopkins University while James and Virginia Stowers donated $985 million to fund the Stowers Medical Institute in Kansas City, Missouri. The article stresses that the controversy surrounding federal funding of stem cell research is misplaced and misleading. The private sector has (IVF research is cited as an example) and will fund biomedical research. The article affirms that "with the private sector taking such great strides, it is futile and self-defeating to go through tortuous efforts to secure government funding" (p. 19); See, S. Fry-Revere and M. Elgin, Public stem cell research funding: Boon or Boondoggle? *Competitive Enterprise Institute: Issue Analysis* 4, September 2008: 1–24; also, J. W. Fossett, Federalism by necessity: State and private support for human embryonic stem cell research, *Rockefeller Institute Policy Brief*, August 9, 2007.
12. G. Q. Daley, Missed opportunities in embryonic stem-cell research, *New England Journal of Medicine* 351, 2004: 627–628; also, George Q. Daley, Testimony by George Q. Daley, MD, PhD. Representing the American Society for Cell Biology before the US. Senate, Commerce Subcommittee on Science, Technology, and Space, September 29, 2004.
13. At least one leading expert in stem cell research, Roger Pedersen, a human embryologist at the University of California, San Francisco (UCSF), announced in July 2001 that he was leaving the United States for Cambridge, England. When he was interviewed about the move, he said: "I wouldn't be able to maximize my potential in the United States, given the restrictions on federal financing and developing cell lines." Another seasoned researcher, also at UCSF, Didier Stanier, eventually left the United States for the Max Plank Institute in Germany. Pedersen is often referenced in books and articles on stem cell research as an example of the consequences of federal funding uncertainty. See, T. Abate, UCSF stem cell expert leaving U.S./Scientist fears that political uncertainty threatens his research, *SFGate* July 17, 2001; also, A. Park, *The Stem Cell Hope: How Stem Cell Medicine Can Change Our Lives*, New York: Hudson Street Books, 2011: 66–67, 139–140; also, N. Wade, Scientists divided on limit of federal stem cell money, *New York Times*, August 16, 2001.
14. N. Wade, Stem cell researchers feel the pull of the Golden State, *New York Times*, May 22, 2005.
15. International Council for Science, The Value of Basic Scientific Research, December 2004.
16. U.S. Senate, Joint Economic Committee, *The Benefits of Medical Research and the Role of the NIH*, May 2000.
17. B. Obama, Remarks on signing an executive order removing barriers to responsible scientific research involving human stem cells and a memorandum on scientific integrity, March 9, 2009, *Public Papers of the Presidents of the United States*, Book 01, March 9, 2009: 199.
18. U.S. Department of Health and Human Services, Centers for Disease Control and Prevention, National Center for Health Statistics, *Health United States*, 2015, Table 14, p. 93.

19. U.S. Department of Health and Human Services, Centers for Disease Control and Prevention, National Center for Health Statistics, *Health United States*, 2015, Table 19, p. 107.
20. S. Simon, Annual report: Cancer deaths in the U.S. continue to decline, *American Cancer Society*, December 16, 2013.
21. PRNewswire, Asterias biotherapeutics announces positive new long-term follow-up results for AST-OPC1, May 24, 2016.
22. ViaCyte, Inc., a company based in San Diego and funded, in part, by CIRM, began recruiting patients in 2014. VC-O1 Combination Product is in the Clinic, at viayte.com/clinical/clinical-trials/.
23. B. D. Colen, Potential diabetes treatment advances, *Harvard Gazette*, January 25, 2016.
24. National Institutes of Health, A Short History of the National Institutes of Health.
25. NIH, Clinical Centers.
26. NIH, National Center for Advancing Translational Sciences.
27. Pfizer, Centers for Therapeutic Innovation. An introduction to the Centers sums up the goal of the Centers as—Translating Leading Science into Clinical Candidates through Networked Collaboration.
28. C. E. Gulbrandsen et al., Legal framework pertaining to research creating or using human embryonic stem cells, in J. Odorico, S.-C. Zhang and R. Pedersen (editors), *Human Embryonic Stem Cells*, New York: Garland Science/BIOS Scientific Publishers, 2005: 332.
29. NSF, Report to the National Science Board on the National Science Foundation's Merit review Process Fiscal Year 2013.
30. https://www.nih.gov/about-nih/what-we-do/budget.
31. Figures from the Federation of American Societies American for Experimental Biology (FASEB), a peak nonprofit organization that advocates for increased funding of basic research. See, FASEB, *Federal Funding for Biomedical and Related Life Sciences Research.*
32. NSF, Report to the National Science Board on the National Science Foundation's Merit review Process Fiscal Year 2013.
33. S. Rockey, Application success rates decline in 2013, *National Institutes of Health Office of Extramural Research Blog*, December 18, 2013; http://nexus.od.nih.gov/all/2013/12/18/application-success-rates-declinein-2013/.
34. Ibid.
35. Ibid.
36. K. Boadi, Erosion of funding for the National Institutes of Health threatens U.S. leadership in biomedical research, *Center for American Progress*, March 25, 2014.
37. American Academy of Arts & Sciences, *Restoring the Foundation: The Vital Role of Research in Preserving the American Dream*, Cambridge, MA, 2014.
38. NIH, Research Portfolio Online Reporting Tools (RePort).
39. American Society for Biochemistry and Molecular Biology, *Unlimited Potential: Vanishing Opportunity*, 2013 Survey.
40. Ibid., p. 6.
41. Ibid., p. 11.
42. NIH, Notice of Criteria for Federal Funding of Research on Existing Human Embryonic Stem Cells and Establishment of NIH Human Embryonic Stem Cell Registry; also, J. A. Johnson and E. D. Williams, Stem cell research, *Congressional Research Service Report for Congress*, Updated August 10, 2005.
43. A. M. Miniño et al., *Deaths: Final Data for 2000*, National Vital Statistics Report 50 Hyattsville, MD: National Center for Health Statistics, September 16, 2002.
44. Alzheimer's Association, 2016 Alzheimer's Disease Facts and Figures.
45. Ibid. The 2016 Report includes a chapter on the personal and financial impact of Alzheimer's disease on families.
46. National Institutes of Health Reform Act of 2006 P.L 109-482, January 15, 2007; 120 Stat. 3675.
47. See the link for Federal Funding by State and District which provides an interactive map of how much in federal research funding the state and district received. Federation of American Societies for Experimental Biology.
48. NIH, Research Portfolio Online Reporting Tools (RePort); https://projectreporter.nih.gov/reporter_searchresults.cfm.

49. J. McKinsey, NIH, Central Budget Office Staff. Telephone Interview by Toni Marzotto, May 12 and 13, 2016.
50. Ibid.
51. We used the NIH RePORT database to search for funding results by fiscal years beginning with FY 2002 for stem cell research. The database returned all the expenditures, which can then be sorted by workshop, research grant, location, etc.
52. Page 12 of the CRS Report lists the human embryonic stem cell lines eligible for use in federal research by the entity with proprietary rights. It also lists those lines that were available. See, J. A. Johnson and E. D. Williams, Stem cell research, *Congressional Research Service Report for Congress*, Updated August 10, 2005.
53. J. Bryant, BresaGen moves slowly to grow stem cell work, *Atlanta Business Chronicle*, November 11, 2002.
54. A. B. Parson, *The Proteus Effect: Stem Cells and Their Promise for Medicine*, Washington, DC: Joseph Henry Press, 2004: 244.
55. Ibid.
56. NIH, NIGMS Center Grants to Explore Stem Cell Biology, September 29, 2003; https://www.nigms.nih.gov/News/results/Pages/HESC.aspx.
57. Ibid.
58. Ibid.
59. Ibid.
60. Ibid.
61. We used the NIH RePORT database to search for funding by year and by category, using the same terms used in the RCDC. From these searches we were able to determine how many grants were awarded, amount, institution, and principle investigator.
62. NIH, Office of Legislative Policy Analysis Legislative Updates, *Pending Legislation—110th Congress.*
63. National Conference of State Legislatures, *Embryonic and Fetal Research Laws*, January 1, 2016.
64. K. Gavin, Five years after Michigan vote on human embryonic stem cells, U-M effort is in full swing, November 14, 2013; http://www.uofmhealth.org/news/archive/201311/stemcell5.
65. Michigan Legislature, State Constitution, Section 27 Human Embryo and Embryonic Stem Cell Research.
66. M. Warnock, *A Question of Life. The Warnock Report on Human Fertilisation and Embryology*, Oxford: Basil Blackwell, 1985. Originally entitled Report of the Committee of Inquiry into Human Fertilisation and Embryology (July 1984).
67. A. D. Levine, State stem cell policy and the geographic preferences of scientists in a contentious emerging field, *Science and Public Policy* 39, August 2012: 530–541.
68. H. B. Alberta et al., Assessing state stem cell programs in the United States: How has state funding affected publication trends? *Cell Stem Cell Forum* 16, February 2016: 115–118.
69. Ibid., p. 116.
70. http://keck.rutgers.edu/research-clinical-trials/research-clinical-trials.
71. C. Y. Johnson, UMass stem cell lab to close, *Boston Globe,* June 28, 2012.
72. CIRM, Tissue Collection for Disease Modeling: RFA 12-02; also, CIRM, hiPSC Derivation: RFA 12-03; also, CIRM, hESC Repository: RFA 12-04.
73. Ibid. Each Request for Application (RFA) contains the same language regarding objectives of the request.
74. Cellular Dynamics International announced that it had been awarded funding to set up stem cell banks by CIRM. See, https://cellulardynamics.com/news-events/news-item/cellular-dynamics-and-coriell-institute-for-medical-research-awarded-multi-million-dollar-grants-from-california-institute-for-regenerative-medicine-to-manufacture-and-bank-stem-cell-lines/.
75. A. B. Parson, *The Proteus Effect: Stem Cells and their Promise for Medicine*, Washington, DC: Joseph Henry Press, 2004: 175.
76. A. Park, *The Stem Cell Hope: How Stem Cell Medicine Can Change Our Lives*, New York: Hudson Street Press, 2011: 135–160.
77. W. J. Clinton, E. O. 12975, Protection of human research subjects and creation of national bioethics advisory commission, *Federal Register* 60(193), October 3, 1995: 52063–52065.

78. U.S. Department of Health and Human Services, National Institutes of Health, *Notice; withdrawal of NIH Guidelines for Research Using Human Pluripotent Stem cells Derived from Human Embryos*, published August 25, 2000, 65 FR 51976, corrected November 21, 2000, 65 FR69951; http://stemcells.nih.gov/staticresources/news/newsArchives/fr14no01-95.htm.
79. T. Abate, UCSF stem cell expert leaving U.S./Scientist fears that political uncertainty threatens his research, *SFGate*, July 17, 2001.
80. A. Zitner, Uncertainty is thwarting stem cell researchers, *Los Angeles Times*, July 16, 2001.
81. Ibid.
82. M. Martin, Davis OKs stem cell research/California is first state to encourage studies, *San Francisco Chronicle, Sacramento Bureau*, September 23, 2002.
83. California Senate Bill 253, *Embryonic Stem Cell Research*; ftp://192.234.213.3/pub/01-02/bill/sen/sb_0251-0300/sb_253_bill_20020922_chaptered.html.
84. Martin, Davis OKs stem cell research/California is first state to encourage studies.
85. This bill introduced by Senator Sam Brownback (R-KS) and Representative Dave Weldon (R-FL) would have banned both publicly and privately funded human cloning based on somatic cell nuclear transfer. The bill would also have banned the importation of any product of human cloning for any purpose.
86. W. W. Gibbs, The California gambit, *Scientific American*, June 27, 2005.
87. W. S. Hwang et al., Evidence of a pluripotent human embryonic stem cell line derived from a cloned blastocyst, *Science* 303, March 12, 2004: 1669–1674; also, N. Wade and C. Sang-Hun, Human cloning was all faked, Koreans report, *New York Times*, January 10, 2006.
88. P. Elias, America lags in stem cell research: Political pressure limits funds for cloning experiments, *NBC News*, updated February 13, 2004.
89. Ibid.
90. A. Park, *The Stem Cell Hope: How Stem Cell Medicine Can Change Our Lives*, New York: Hudson Street Press, 2011: 138.
91. Ibid., p. 139.
92. California Stem Cell Research & Cures Initiative Webpage is archived by the University of California at Los Angeles; http://digital.library.ucla.edu/websites/2004_996_027/index.htm.
93. N. Vogel, Investors pour millions into Prop. 71 race, *Los Angeles Times*, October 18, 2004.
94. M. Hiltzik, Did the California stem cell program promise miracle cures? *Los Angeles Times*, May 30, 2012.
95. Vogel, Investors Pour Millions into Prop. 71 Race.
96. UC Hastings College of Law, *California Ballot Measures*; www.library.uchastings.edu/research/ballots/index.php.
97. A. Pollock, California stem cell research is upheld by appeals court, *New York Times*, February 27, 2007.
98. A. Pollack, Trial over California stem cell research ends, *New York Times*, March 3, 2006.
99. A. Park, *The Stem Cell Hope: How Stem Cell Medicine Can Change Our Lives*, pp. 156–159; also, D. G. Riordan, Perspective: Research funding via direct democracy: Is it good for science? *Issues in Science and Technology* 24, Summer 2008.
100. CIRM, *Our Governing Board*; https://www.cirm.ca.gov/board-and-meetings/board.
101. M. Hiltzik, Stem cell agency not doing enough to avoid conflict to interest, *Los Angeles Times*, January 29, 2013.
102. Ibid.
103. S. M. Lee, Stem cell agency acts to prevent "potential conflict of interest," *San Francisco Chronicle*, Wednesday, July 9, 2014.
104. CIRM, *Search CIRM Grants*; https://www.cirm.ca.gov/grants.
105. Hiltzik, Stem cell agency not doing enough to avoid conflict to interest.
106. G. W. Bush, Message to the House of Representatives, returning without approval to the Senate the Stem Cell Research Enhancement Act of 2005, *Public Papers of the Presidents of the United States, G. W. Bush*, Book 02, July 19, 2006: 1425.
107. NIH, *Finding a Clinical Trial*.
108. California Institute for Regenerative Medicine, *Beyond CIRM 2.0*: *Proposed Strategic Plan 2016 and Beyond*.
109. CIRM, *Stem Cell Agency okays $150 "powerhouse,"* April 11, 2016.

110. See, California Institute for Regenerative Medicine, *Beyond CIRM 2.0: Proposed Strategic Plan 2016 and Beyond*; also, National Institutes of Health, *NIH-Wide Strategic Plan: Fiscal Years 20116-2020: Turning Discovery into Health.*
111. National Institutes of Health, ibid.
112. Ibid.
113. Ibid.
114. CIRM, *The Stem Cellar: The official blog of CIRM, California's Stem Cell Agency*, June 28, 2016.
115. National Institutes of Health, *Mission and Goals.*
116. S. Fry-Revere and M. Elgin, Public stem cell research: Boon or boondoggle? *Competitive Enterprise Institute, Issue Analysis* 4, 2008; also, S. Fry-Revere, No tax money for stem cells, *Los Angeles Times*, November 28, 2006.
117. J. W. Fossett, Federalism by necessity: State and private support for human embryonic stem cell research, Rockefeller Institute Policy Brief, August 9, 2007; also, D. Spar and A. Harrington, Selling stem cell science: How markets drive law along the technological frontier, *American Journal of Law & Medicine* 33, 2007: 541–565; also, J. Jacoby, Let the private sector fund stem-cell research, *Boston Globe*, August 29, 2010.
118. S. Fry-Revere and M. Elgin, Public stem cell research: Boon or boondoggle? p. 12.
119. Ibid, p. 19.
120. U.S. Centers for Disease Control and Prevention, *The Fertility Clinic Success Rate and Certification Act*. Pub. L. 102-93 106 Stat. 3146, October 24, 1992.
121. WETA, *The Pros and Cons of IVF. The American Experience*; https://www.pbs.org/wgbh/americanexperience/features/general-article/babies-pros-and-cons/.
122. D. Spar and A. Harrington, Selling stem cell science: How markets drive law along the technological frontier.
123. A. D. Levine and L. E. Wolf, The roles and responsibilities of physicians in patients' decisions about unproven stem cell therapies, *Journal of Law, Medicine, & Ethics* 40, Spring 2012: 122–134; also, D. Lau et al., Stem cell clinics online: The direct-to-consumer portrayal of stem cell medicine, *Cell Stem Cell* 3, December 4, 2008: 591–594.
124. L. Turner and P. Kroepfler, Selling stem cells in the USA: Assessing the direct-to-consumer industry, *Cell Stem Cell* 18, June 30, 2016.
125. L. McGinley, Unregulated stem-cell clinics proliferate across the U.S., *Washington Post* A3, July 1, 2016.
126. Ibid.
127. Ibid.
128. Ibid.
129. International Society for Stem Cell Research, *Guideline for Stem Cell Research and Clinical Translation*, May 12, 2016.
130. T. Caulfield et al., Confronting stem cell hype, *Science* 352, May 13, 2016: 776–777.
131. International Society for Stem Cell Research, *Guideline for Stem Cell Research and Clinical Translation*, pp. 28–29.
132. J. Victory, Journalists: 9 tips to combat stem cell hype in your news stories, *Health News Review*, June 27, 2016.
133. Ibid.
134. B. Obama, Remarks on signing an executive order removing barriers to responsible scientific research involving human stem cells and a memorandum on scientific integrity, March 9, 2009, *Public Papers of the Presidents of the United States*, Book 01, March 9, 2009:199.
135. Cambridge Academic Content Dictionary; www.dictionary.cambridge.org/us/dictionary/english/brain-drain.
136. International Society for Stem Cell Research, *About ISSCR*; www.isscr.org/home/about-us/about-the-isscr.
137. A. D. Levine, Policy uncertainty and the conduct of stem cell research, *Cell Stem Cell* 8, February 4, 2011: 132–135.
138. A. Park, *The Stem Cell Hope: How Stem Cell Medicine Can Change Our Lives*, New York: Hudson Street Books, 2011: 66–68; also, E. Herold, *Stem Cell Wars: Inside Stories from the Frontlines*, New York: Palgrave Macmillan, 2006.
139. A. Levine, Trends in the geographic distribution of human embryonic stem-cell research, *Politics and the Life Sciences* 23, September 14, 2005: 40–44.

140. C. Vaughn, Back in the Bay Area: Sabbatical draws stem cell expert Pedersen to Stanford, *Stanford Medicine News Center*, November 19, 2012.
141. A. Zitner, Uncertainty is thwarting stem cell researchers, *Los Angeles Times*, July 16, 2001.
142. Ibid.
143. R. Trager, U.S. funding boost threat to EU science? *Royal Society of Chemistry News*, April 6, 2009.
144. E. Herold, *Stem Cell Wars: Inside Stories from the Frontlines*, New York: Palgrave Macmillan, 2006: 204.
145. Editorial, Singapore's salad days are over, *Nature* 468, December 2010.
146. Hudson Institute of Medical Research, IVF pioneer, Alan Trounson returns to the Institute; http://hudson.org.au/news-events/latest-news/ivf-pioneer-alan-trounson-returns-to-the-institute.
147. M. Santora, Researchers on stem cells are making do, and hoping, *New York Times*, September 17, 2003.
148. T. H. Maugh II, Xiangzhong "jERRY" Yang dies at 49; leading researcher in cloning technology, *Los Angeles Times*, February 10, 2009.
149. K. Birmingham, Europe fragmented over embryonic stem cell research, *Journal of Clinical Investigation* 112, August 15, 2003: 458.
150. Ibid.
151. EuroStemCell, About Us; www.eurostemcell.org/about.
152. A. D. Levine, Policy uncertainty and the conduct of stem cell research, *Cell Stem Cell* 8, February 2011: 132–135.
153. C. Vaughn, Stanford creates first PhD program in stem cell science, *Stanford Medicine News Center*, November 19, 2011.
154. Research!America, Polls and Publications; www.researchamerica.org/polls-and-publications. The 2005 poll, Americans Speak Out on Stem Cell Research, was conducted in June 2005 by the Charlton Research Company. The 2012 poll was conducted by JZ Analytics in September 2012. The 2016 poll was conducted by Zogby Analytics in January 2016.
155. G. Vogel, Breakthrough of the year: Capturing the promise of youth, *Science Magazine* 286, December 17, 1999: 2238–2239.
156. F. Golden and D. Thompson, Cellular biology: Stem winder, *Time* 158, August 20, 2001: 26–29.
157. E. Herold, *Stem Cell Wars: Inside Stories from the Frontlines*, New York: Palgrave Macmillan, 2006: 203.
158. C. Fox, *Cell of Cells: The Global Race to Capture and Control the Stem Cell*, New York: Norton, 2007: 46–47.
159. S. Len, South Korea, with renowned scientists, jolts field and revives debate, *New York Times*, February 13, 2004.
160. Editorial, Time to put U.S. at the forefront of promising research, *USA Today*, May 24, 2005.
161. K. Human, U.S. researchers recruited labs in China, South Korea, and England are attracting scientists frustrated by the lack of stem cell funding, *Denver Post*, May 25, 2005.
162. As quoted in C. Fox, *Cell of Cells: The Global Race to Capture and Control the Stem Cell*, p. 419.
163. P. Elias, America lags in stem cell research: Political pressure limits funds for cloning experiments.
164. As quoted in Cynthia Fox, p. 419.
165. N. Wade, Stem cell work earns Nobel Prize, *New York Times*, October 8, 2012; Also, C. Mummery, and others, *Stem Cells: Scientific Facts and Fiction*, London: Elsevier, 2011: 84–85.
166. R. Stengel, The Time 100 Team, *Time* 171, May 12, 2008: 96.
167. J. A. Johnson and E. D. Williams, *Stem Cell Research: Federal Research Funding and Oversight*, *Congressional Research Service Report*, RL 33540, April 18, 2007.
168. J. Chakma et al., Asia's ascent—Global trend in biomedical R&D expenditures, *New England Journal of Medicine* 370, January 2, 2014: 3–6.

6 Ethical dilemmas

Always changing

No book on stem cell research would be complete without a chapter on the ethical issues that have swirled and continue to swirl around research using human tissues and, particularly, stem cells. It is important for the reader to understand the variety of perspectives that shape opinions on this research. Key ethical issues include: beliefs on when human life begins; the deep-seated beliefs that shape those opinions; and how to regulate this research to ensure proper informed consent, appropriate use of public funding, protection of participants, and equitable access to results. There is no consensus on any of these topics, but they are critical to consider as we move forward into an era where the results of stem cell research will be expanding.

The following is a brief overview of common perspectives on ethics in health care and research. There is general agreement that the health care system's primary duty is to provide good patient care and protect patients from harm. The field of ethics overall includes personal moral standards based on culture, religion, family, and society, as well as professional and legal categories by which appropriately ethical behavior is measured. Health care ethics is the area specifically related to the practice of medicine, including patients, caregivers, health care professionals, administrators, researchers, and the general public. Bioethics is a type of applied health care ethics focusing primarily on ethical choices stemming from advances in medical care and research. There is considerable confusion about the boundaries of ethics and bioethics, particularly in a time when medical research seems to be moving quickly and the general public is eagerly awaiting "magical cures."

Commonly used ethical theories

While some authors may use slightly different terminology, the primary categories of ethical theories used in health care are based on views about: results, duty, virtue, justice, and rights. Of course, the general public is unlikely to use that specific terminology, but these perspectives do shape their beliefs. As an example, when President George W. Bush spoke about stem cell research in 2001, he emphasized his moral concern over what scientists were doing, while also recognizing patients' and advocates' eagerness for cures. Similarly, President Barack Obama recognized the need for federal support for research in his Executive Order in 2009 (see Chapter 3). Both of these speeches touched on ethical expectations, without getting into detailed theories. Some of the terms used in academic discussions include:

1. Utilitarianism, or *Teleology*, which comes from the Greek word "telos" for "end or result." "The end justifies the means" in this approach, meaning that the rightness or wrongness of an act depends on how much happiness it creates compared to another choice. This approach is most commonly linked with John Stuart Mill, and includes: (a) act utility = each decision one at a time, and (b) rule utility = patterns of decisions with a cost/benefit balance over all of them.[1] In terms of stem cell research, it would measure the results obtained, not necessarily beliefs or laws.

2. Duty-based ethics, or *Deontology*, which comes from the Greek word "deon" for "duty," is linked to Kant. In this approach, there is an independent right and wrong which leads to a duty to one's fellow humans. One must treat people as the end of actions, not just as means to your success. Goodwill and a sense of duty are the key. The Golden Rule of "do unto others as you would have them do unto you" is central to this approach.[2] Of course, one's view of "right and wrong" can vary depending on cultural or religious background, affecting perspectives on research of all sorts. In terms of stem cells, a key factor is determining when an entity is a "person," deserving respect for its dignity.[3]
3. Virtue-based ethics is based on the view that a rational man can discern what is good and evil, and will choose to do good, in the pursuit of happiness. A series of philosophers from Aristotle to Aquinas described this view, but there is concern about being too trusting that others will always produce good results.[4] Leon Kass, while leading the President's Council on Bioethics (under President Bush), called for principled reasoning based on "the broader plane of human procreation and human healing, with their deeper meaning," rather than trying to seek compromise.[5]
4. Justice-based ethics, or *Social Contract theory*, is based on John Rawls' theory that rational people will always choose to do good, but that we have a "veil of ignorance" and can't always know exactly what the good *is*. Therefore, we should always protect the least well off, since we could be in that position ourselves.[6] This concern about "potential" humans becomes particularly central when we examine in vitro fertilization (IVF) by-products. The National Bioethics Advisory Commission (under President Bill Clinton) maintained that public deliberation was central to its process and concluded with the belief that human embryos should be accorded respect and not destroyed "arbitrarily," but that the use of products left over from IVF treatments should be acceptable.[7]
5. Rights-based ethics is based on the belief that all people have individual rights, purely because they are human, and that there is a collective responsibility to protect them. This belief is central to many aspects of American democracy, such as freedom of speech. In health care, it can include equal access to protections (such as under the Occupational Safety and Health Act), but it can be difficult in situations with limited resources.[8] A key unsolved question here is: when does "personhood" begin? We will continue to wrestle with that issue in the next section.

Beliefs on when human life begins

Beliefs on when human life begins have shifted over time. As pointed out in Chapter 2, this is an area with no universal definition. It is also an area fraught with religious, emotional, and legal challenges. The major definitions are described here and elaborated on in the next section. Key possibilities include:

1. When an egg is created, making it possible for pregnancy to occur. The Catholic Church has, as a tradition, opposed contraception, because it prevents the creation of a human being. There is no agreement within Christianity about whether an unfertilized egg is actually a "person" with a soul, but there is a strong belief that it has the potential to become a person.[9]
2. When an egg is fertilized and becomes a zygote, or a potential "embryo," even if its cells are not yet differentiated. This perspective, supported by the Catholic Church, argues that a fertilized egg deserves unconditional protection, whether it has an "immortal soul" or not.[10]
3. When the cells in a blastocyst are no longer "pluripotent" and it begins to develop into an actual embryo (day 14 after fertilization, according to the National Institutes of Health (NIH) and the Warnock Commission).[11]

4. When an embryo/fetus is "viable," or capable of survival outside of the womb. This also varies depending on the perspective taken and, as medicine has advanced, the former method of using "trimesters" of pregnancy is less applicable.

What shapes positions on human life?

Beliefs vary across the spectrum of religions and cultures, shaping positions and affecting laws. As an example, although the Catholic Church does not approve of contraception, it is legal in most countries and in all American states for adults. Similarly, abortion is legally defined as the termination of a pregnancy before the fetus is "viable," although the laws and ethical beliefs about it have varied over time. In the United States, a key Supreme Court case was *Roe v. Wade* decided in 1973. This case decided that a Texas law prohibiting abortions not necessary to save the mother's life was a violation of a woman's right to privacy under the Fourteenth Amendment. In the decision, the Supreme Court adopted a process linked to the trimesters of pregnancy and the viability of the fetus. It proclaimed abortions prior to viability to be a decision between the woman and her physician, but after viability, the state would have a compelling interest in saving the life of the child, unless to do so would harm the life or health of the mother.[12]

Since that time, medical progress has made it more difficult to tie viability to trimesters and court cases about abortion laws continue to arise, including the *Planned Parenthood of Southeastern Pennsylvania v. Casey* case in 1992, in which the Supreme Court affirmed the mother's right to terminate her pregnancy before the point of "viability," while rejecting the trimester model.[13] In a 2016 case, *Whole Woman's Health v. Hellerstedt*, the Supreme Court decided that Texas cannot place restrictions on the delivery of abortion services that create an "undue burden" on women seeking an abortion. This will likely affect access in other states as well, with a flood of restrictive laws and court cases underway.[14]

While human embryonic stem cells (hESC) are no longer derived from aborted fetuses, at this point they can come from umbilical cords, and some belief systems also regard destroying unused IVF products as "abortion," so the discussions can be complicated. There has been no universal agreement among religious traditions about when an embryo becomes a "human being." In the following section, we will briefly explore the perspectives of religious groups.

Religious perspectives

Christianity

Although politicians tend to cite the views that support their own beliefs, there is a wide variation in Christian beliefs about embryonic research. While some Christian traditions focus on the point of fertilization as creating "personhood," others maintain that the soul enters the embryo at some later point (making it "human"), perhaps at the stage of implantation. In the Roman Catholic perspective, the key point is that embryos are "potential" human beings, to be protected from the beginning, hence their opposition to contraception and abortion. However, the Catholic Church, along with multiple Protestant denominations, has been engaged in ongoing discussions about exactly when the fertilized egg becomes an individual human being. While a series of proclamations has come from these bodies about the issue, there is no unified agreement. The Vatican has hosted a series of conferences on stem cell research, promoting the use of therapies based on adult stem cells. Under Pope Benedict XVI, in 2011 the use of adult stem cells to heal chronic illnesses was encouraged. In 2016, Pope Francis spoke about the need for empathy for all of those suffering from rare diseases, ensuring that each person, "regardless of culture, social standing, or religious beliefs," has access to needed care, as well as declaring ongoing support for adult stem cell research, as long as it "safeguards human life and the dignity of the person."[15]

Judaism

Jewish traditions have generally held that human embryos may be used for potentially lifesaving research. While there is no single voice speaking for their tradition, Jewish scholars have generally held that until the fortieth day after conception, embryos are "as if they were simply water." The primary consideration for this research is whether it will save existing lives. As we discussed in Chapter 2, there have been multiple debates about the morality of conceiving a child in order to provide needed stem cells to heal a sibling's genetic disorder. In the Molly Nash case, the family protested that their Jewish beliefs did not conflict with using IVF to conceive her brother, whose umbilical cord blood stem cells did, indeed, solve Molly's condition, and both of them survived the process.[16]

Islam

Islamic scholars also agree that the embryo does not have "personhood" until later than the point at which stem cells would be derived. Depending on the particular sect within Islam, this can be anywhere from 40 to 120 days after conception. In any case, they all agree that it occurs later than the point at which cells are no longer pluripotent (5 days after conception). The key factor is whether the research on stem cells is undertaken in order to provide treatment to the sick and suffering, as the Koran would encourage.[17]

Other major religions

While Hindu beliefs are adamantly opposed to abortion, there are a variety of traditions about when "personhood" begins, which might allow stem cell research if motivated by compassion and aimed at therapeutic uses. There is debate among scholars of Buddhism about at what point the embryo is owed protection. Compassion is also a central thread in this tradition, leading to perspectives stating that the use of embryos in stem cell research would be acceptable if done in the service of others.[18]

Overall, a wide variety of religious beliefs and perspectives do agree on the critical importance of protecting human life and they are all exploring exactly how stem cell research fits into this central value. However, the science has been allowed to proceed in many countries with a wide variety of religious views. In the United States, for instance, the privately funded research sector had no religious/political restraints on it despite adamant resistance to stem cell research among some sectors, slowing public funding for the field.

National attention to bioethical concerns and research progress

Since 1974, there have been a series of national commissions in the United States exploring various aspects of medical ethics. While some were created in response to the revelation of unethical treatment of research subjects, most were trying to be proactive about the evolving nature of biological research, its possible impact on human safety, and the appropriate role of federal funding and regulation of science. The brief descriptions that follow outline each group's major focus, its findings, and the impact on science (intended or unintended). Each one was deeply affected by the politics of the time and their methods of operation, as well as their reports and recommendations, reflect that.

The first public national body on medical ethics in the United States was the National Commission for the Protection of Human Subjects of Biomedical and Behavioral Research. It was created by Congress in 1974, after a series of disclosures about research conducted with unethical treatment of human subjects (beginning with the Tuskegee Study). It produced multiple reports about different aspects of human research, leading up to the Belmont Report in 1978 and the creation of the human subjects review process in the United States. It operated under the Department of Health, Education, and Welfare (now the Department of Health and Human Services). The Office of Human Research Protection and the Institutional Review Board (IRB) process have existed ever since (with varying names), to ensure that organizations receiving federal funding follow the "Common Rule," a Federal law aimed at protecting all human research participants.[19]

After Louise Brown's birth via IVF in 1978, Congress quickly convened a federal ethics advisory board (EAB) to address whether the U. S. government should fund any research on in vitro fertilization and, if so, what would be acceptable. The board came back with a report in May 1979 that supported a wide range of studies on human embryos, but required that these studies would have to be approved by the EAB, as well as their local review boards, before they could receive any federal funding. The report immediately caused an avalanche of letters to the Department of Health and Human Services based on moral and religious principles hostile to any type of IVF research. In quick succession, Secretary Califano was fired and the EAB's charter was allowed to expire. Without it, no IVF research could be reviewed and with President Ronald Reagan's election in 1980 the possibility of federal funding lapsed, leaving the field, as well as review of IVF and other types of research, to private sector support and state-level funding for over a decade.[20]

After Bill Clinton was elected president, the National Institutes of Health Revitalization Act was passed in 1993, allowing the NIH to fund human embryonic research again and eliminating the requirement that a federal Ethics Advisory Board had to approve any research. However, the Administration wanted advice from medical, ethical, and research experts on how to approach this ethically complicated area. So the Human Embryo Research Panel (HERP) was created to review a wide range of information, including the perspectives of various religions on the beginning of "personhood," reports prepared by other countries (including the Warnock report in England), and views on when it would be acceptable to use excess IVF embryos or to create human embryos for research purposes only. When the report was released in 1994, there was tremendous opposition to funding the creation of embryos for research purposes.[21] By 1996, the Dickey–Wicker amendment had been attached by Congress to the NIH appropriations bill, ensuring that no federal money could be spent on studies in which an embryo is destroyed, including studies on human embryonic stem cells (hESCs).[22] As discussed in Chapter 3, this amendment has had an enduring effect on the field.

Broadening the perspective from "human embryo research," the National Bioethics Advisory Commission (NBAC) was established by President Clinton, using an Executive Order in late 1995, to "provide advice and make recommendations to the National Science and Technology Council (in the White House), other appropriate entities and the public, on bioethical issues arising from research on human biology and behavior, including the clinical applications of that research."[23] While this commission built on the previous work of the National Commission for the Protection of Human Subjects of Biomedical and Behavioral Research, it also expanded its priorities to include "issues in the management and use of genetic information, including, but not limited to, human gene patenting."[24] There were two subcommittees, one focused on finding gaps and updating regulations for human subjects research and the other (the Genetics Subcommittee) examining issues that were rapidly arising with genetic research. In the course of the NBAC's existence, it produced reports on ethical issues in human stem cell research, human biological materials, and cloning human beings.[25]

The President's Council on Bioethics (PCBE) was a group appointed by President Bush under Executive Order 13237 in November 2001, after his August speech on allowing public funds to be spent again for stem cell research on existing cell lines. The Council was created to "advise the President on bioethical issues that may emerge as a consequence of advances in biomedical science and technology" and to replace the National Bioethics Advisory Commission. The president directly appointed the members of the Council and it was accused by former members of being set up to justify Bush's positions on stem cell research and abortion as well as to "wrap political and religious agendas in the guise of dignity" from a conservative perspective rather than to provide actual policy guidance.[26] It produced reports on monitoring stem cell research, alternative sources of pluripotent cells, and the responsible regulation of new biotechnologies. After the election of President Obama, it was replaced by the Presidential Commission for the Study of Bioethical Issues, created in November 2009 by Executive Order 13521.[27]

The Presidential Commission for the Study of Bioethical Issues is an advisory panel made up of members drawn from the fields of medicine, science, ethics, religion, law, and engineering. It was created to advise the president on bioethical issues arising from advances in biomedicine and related areas in science and technology. It seeks to identify and promote policies and practices that ensure scientific research, health care delivery, and technological innovations are conducted in a socially and ethically responsible manner. It has produced a series of reports on a wide variety of topics, including: federal oversight of synthetic biology, improving human subjects research protection, whole genome sequencing, and the interaction of ethics with current issues (including Ebola and the development of clustered regularly interspaced short palindromic repeats (CRISPR) technology).[28]

Each of these committees was created in response to the political and research situation at the time. While all of them sought to encourage appropriate standards, and to balance beliefs and a concern for public safety, their structures and reporting mechanisms varied widely. The question of proper research behaviors does not go away, but at the same time there is strong public pressure to expedite approval of treatments and flawed oversight of clinics claiming to have "cures," as we discussed in Chapter 5. What is the "proper" level of regulation in this area?

Ongoing and emerging concerns

While the field of genetic engineering is still evolving, there have been some useful therapies developed and great hope remains for future results, particularly from stem cell research. There is an ethical concern that gene therapy might be used for some unnecessary reasons (to make a person taller, perhaps), but also the hope that genetic diseases might be corrected. A technique called CRISPR-Cas 9, which can potentially reprogram human embryonic deoxyribonucleic acid (DNA), is being studied intensely.[29] It has raised awareness that there are no U.S. laws governing the type of research done on embryos with private money and that, if it were done on stem cells, the revised traits could potentially be passed on to future generations.[30]

A major concern in this area is whether practitioners are following the appropriate ethical standards in providing care to patients. There are multiple legal and ethical structures in place aimed at the primary goals of promoting professional competence, while protecting patients' rights. A central question for stem cell research (tied once again to the determination of "personhood") is whether the rights of a potential "person" are equal to those of the living person making decisions for oneself or family members. Another key area of concern is how best to balance the rights of different groups of individuals. In her thought-provoking book, *People's Science: Bodies and Rights on the Stem Cell Frontier,* Ruha Benjamin discusses the conflicting perspectives on the regulation of stem cell research among different groups and the struggles, particularly in California, to ensure that publicly funded research is used not only to benefit society as a whole, but also to benefit those in most need of the results. She discusses the need for social responsibility and for the participation of potential recipients of treatments in the choices about their development.[31] Other authors have also explored this question in depth, citing efforts (particularly in Europe) to include patient advocacy movements in the regulation of stem cell research and the allocation of research funding.[32]

In December 2015, there was an international summit on human gene editing hosted by the National Academies of Sciences and Medicine. The result was a call for an international forum to discuss the science of human gene editing; the clinical, ethical, legal, and social implications of the use of these techniques in research and medicine; and how to promote research coordination among nations. Their initial perspectives were:

1. Basic and preclinical research is needed, but should proceed with appropriate legal and ethical oversight. At this early stage of research, they concluded that "If, in the process of research, early human embryos or germline cells undergo gene editing, the modified cells should not be used to establish a pregnancy."[33]
2. Clinical use of gene editing in cells whose genomes are not transmitted to the next generation does require an understanding of the risks and benefits of each proposed modification, but since these treatments only affect the person receiving them, they can be evaluated under existing regulatory frameworks.[34]
3. Clinical use of gene editing in embryos, where the changes might be passed on to future generations, requires much more intensive scrutiny, and "it would be irresponsible to proceed with any clinical use of germline editing unless and until (i) the relevant safety and efficacy issues have been resolved … and (ii) there is broad societal consensus about the appropriateness of the proposed application." Also, any clinical use should proceed under "appropriate regulatory oversight," which did not yet exist.[35]

These guidelines were shaped by five guiding principles, setting standards for the ethical development of stem cell interventions. These principles are: integrity of the research enterprise, primacy of patient welfare, respect for research participants, transparency, and social justice. While these standards have been cited repeatedly for decades, there is always a need to reiterate them in the development of new treatments and (potentially) regulations.[36]

Hence, the National Academy of Sciences (NAS) and the National Academy of Medicine's Committee on Human Gene Editing: Scientific, Medical, and Ethical Concerns was formed and charged with developing a comprehensive report after numerous meetings and solicitation of input from scientists, ethicists, clinicians, and the public. Their list of topics covered as of July 2016 included: models of public engagement in policy, perspectives from affected communities, developing potential therapeutic applications, principles underlying governance (including ethics and interest group views), international perspectives, race and genetics in history, and the interaction of ethics, morality, and public policy in germline editing.[37]

Another controversy that has arisen is about the "ownership" of unused IVF embryos. Since the process of in vitro fertilization involves the creation of multiple fertilized embryos and they can be frozen and cryopreserved indefinitely, there are major issues around who decides the fate of the embryo and whether it should be considered to have rights as a "person." In their work, Cohen and Adashi describe the controversies and common threads in multiple lawsuits on this issue, with particular attention to the wide variation among state laws and case results. In particular, they examine the Missouri case of *McQueen v. Gadberry*, in which a conservative group argued that the "personhood" of the frozen IVF embryo means that its "best interests" must always be protected, leading to debates over rules requiring that embryos must always be preserved or implanted and not destroyed or used for research. They lay out a proposition for a national set of uniform rules, ideally to be passed at the national level, but possibly done on a state-by-state basis if necessary. The five key elements would be: clear differentiation between informed consent and agreements for embryo disposition through standardized forms; a requirement for an embryo disposition agreement prior to embryo cryopreservation; a rule that embryo disposition agreements should ordinarily be binding, even if one party later changes his or her mind; recognition that embryo disposition agreements should not impose legal parenthood obligations on the objecting party; and consideration of special rules for loss of fertility. While a uniform federal law on the subject would be ideal, struggles over abortion rights and restrictions will likely prevent its enactment and the "personhood" movement will use any effort to regulate embryo disposition as another opportunity to protect their "must implant or must-make-available-for-implantation rules."[38]

Key remaining dilemmas

In their thoughtful analysis of ethical issues in stem cell research, Lo and Parham provide a summary of the types of issues that arise at different phases of research. Informed and voluntary consent is a critical topic, particularly with the donation of embryos for research and for participants in early stage clinical trials. The "therapeutic misconception" that one will personally benefit from clinical research is hard to overcome, particularly in areas receiving high levels of press coverage. Researchers need to clearly discuss all aspects of the process, including clarifying the participants' beliefs about the moral status of the cells being used, and to ensure that the participants have a realistic understanding of the trial itself.[39]

Standardized regulation of this research is another area of great concern. As pointed out throughout this text, state laws vary and while national committees have strongly suggested the appointment of Stem Cell Research Oversight Committees (SCROs), they are *not required* unless the research is federally funded. In the United States and other countries, there has been extensive reconsideration of methods for balancing time and resources needed for the research itself with that needed to meet regulatory requirements. Again, there is no consistent agreement on the best approach to use and considerable argument about whether the existing models are most efficacious and whether they are enforced.[40]

Equitable access to potential treatments derived from stem cells also poses multiple conundrums. While there are a wide variety of interventions being marketed in the United States and internationally, there is a mixed response to their ethical acceptability. Some authors cite surveys about the varying connections between religious beliefs and willingness to allow or forbid stem cell research in countries across the globe, while pointing to a growing support for it in America.[41]

Others examine a wider array of beliefs and approaches in other countries, including China, where "stem cell tourism" can be seen as a form of "bionetworking" or entrepreneurial science that incorporates local beliefs, potentially to the benefit of the patient, rather than being purely unethical and greedy.[42] There is also a great deal of concern about unproven treatments being marketed to desperate patients, potentially causing more harm than good. At a point in time when there is international discussion about how best to provide governance for the future of genetic/stem cell research, these questions will continue to boil.

Meanwhile, the Presidential Commission for the Study of Bioethical Issues (PCSBI) published *Bioethics for Every Generation: Deliberation and Education in Health, Science, and Technology* in May 2016. This document is explicitly aimed at improving Americans' knowledge and their ability to deliberate about possible solutions to bioethical questions as new research and treatment options continue to unfold. It focuses on the need for collaborative decision making, particularly when decisions must be made about the distribution of scarce resources or the restriction of individuals' activities in order to ensure the community's well-being. It provides materials about how best to conduct open discussion and debate as well as the education of individuals at all ages about bioethical concerns.[43]

Summary

As we move forward in this area, ethical concerns will continue to arise. There is no one uniform answer to the issue of "personhood" or to the question of where to spend scarce research and treatment dollars in areas of critical health importance to the population. Surveys continue to find the general public is wary of such topics as "gene editing" to produce healthier babies, with wide variation in views about the appropriateness of "meddling with nature," often shaped by respondents' religious beliefs. The Pew Research Center's 2016 survey of over 4700 people found that while respondents saw science as having a mostly positive effect on society (67%), they were very torn over the potential impact of genetic editing.[44] Hopefully, wider awareness of some of the ethical dilemmas involved will increase the public's willingness to engage in careful democratic deliberation, rather than forcing one perspective to trump all others.

Additional readings

Original sources and other scholarly readings

Scholarly research requires reading the original source as well as relevant commentaries. Here is a link to the last meeting of the NAS Committee on Human Gene Editing, along with other sources:

1. National Academy of Sciences and National Academy of Medicine Committee on Human Gene Editing Consensus Study Meeting #4 (July 12, 2016).
2. http://nationalacademies.org/gene-editing/consensus-study/meetings/?utm_source=Human+Gene-Editing+Initiative&utm_campaign=a17c6bddc8-Human_Gene_Editing_July12_slides-posted_8-1-2016&utm_medium=email&utm_term=0_a2539fe65c-a17c6bddc8-278769813&mc_cid=a17c6bddc8&mc_eid=5502d3f645.
3. Pope: Ethical medical research requires morality, Safeguards human life, CNA/EWTN News, April 29, 2016. http://www.ncregister.com/site/print_article/49316.
4. B. Lo and L. Parham, Ethical issues in stem cell research, *Endocrine Review* 30, May, 2009: 204–213.

Secondary analysis and news articles

1. R. Stein, NIH plans to lift ban on research funds for part-human, part-animal embryos, *National Public Radio*, August 4, 2016. (http://npr.org/sections/health-shots/2016/08/04/488387729/nih-pland-to-lift-ban-on-research-funds-for-part-human-part-animal-embryos.)

Critical thinking activities

1. In this chapter, we have focused on the ways in which "personhood" is determined. At what point in development does a cell/embryo/potential person have "rights" that need to be protected and what shapes beliefs about those rights? After viewing the statement by Ronald Cole-Turner in the National Academy of Sciences link above and researching three religions' "take" on this question, write a five-page paper on the key factors influencing these beliefs. Be sure to examine the ways in which these perspectives have (or have not) shifted over time and the likely direction in which they will move in the future.
2. One of the most controversial areas of discussion about stem cell research has been the use of public funds to support it and the lack of adequate regulation of this work. In previous chapters, we have outlined the distinctions between research on hESC and induced pluripotent stem cells (iPSC) and the ways in which beliefs about them have shaped public awareness and funding. Which ethical theories outlined in this chapter would and/or would not support public funding of one or the other of these two types of research and why? What are the key ethical/moral perspectives that have been cited by politicians in taking stands on them? (You can go back to the links in previous chapters to speeches by George Bush, Bill Clinton, and Barack Obama.) Create a table with pros and cons for public funding, using hESC and iPSC as categories, and list the key ethical reasons given in each category by each of those presidents. Then, write a proposal for a national process for regulation of this research to ensure the proper use of public funds and appropriate protections for research participants and future patients.

3. Worry about the potential science fiction aspect of stem cell research arises often. Which is of more concern ethically: editing human DNA to change characteristics for generations to come or using stem cell therapies to heal a given person's conditions without any assurance of what it might mean for future generations? Which groups would argue for and against these efforts and why? What are the major fears raised in these discussions? Start with the link above from NPR and other recent articles about CRISPR and emerging topics in the field. Write a three-page paper outlining the plusses and minuses of using stem cell research for "enhancement" of human beings, rather than only seeking "treatments" for dangerous medical conditions.

Notes

1. B. F. Fremgen, *Medical Law and Ethics*, 5th ed., Boston, MA: Pearson, 2016: 10–11.
2. Ibid., 11–12.
3. B. A. Manninen, Respecting human embryos within stem cell research: Seeking harmony, in L. Gruen, L. Grabel, and P. Singer (editors), *Stem Cell Research: The Ethical Issues*, Malden, MA: Blackwell, 2007: 94–95.
4. Fremgen, 12–13.
5. The President's Council on Bioethics, *Human Cloning and Human Dignity: An Ethical Inquiry*, Washington, DC: 2002: ix.
6. Fremgen, 12.
7. The National Bioethics Advisory Commission, *Ethical Issues in Human Stem Cell Research*, Rockville, MD: U.S. Government Printing Office, 1999: 51.
8. Fremgen, 11.
9. J. Maienschein, The language really matters, in M. Ruse and C. A. Pynes (editors), *The Stem Cell Controversy: Debating the Issues*, 2nd ed., Amherst, NY: Prometheus Books, 2006: 42.
10. T. Pcholczyk, The Catholic Church and stem cell research, in D. L. Kleinman et al. (editors), *Controversies in Science & Technology: From Climate to Chromosomes*, Vol. 2, New Rochelle, NY: Mary Ann Liebert, Inc., 2008: 71.
11. M. Warnock, *A Question of Life. The Warnock Report on Human Fertilisation and Embryology*, Oxford: Basil Blackwell, 1985. Originally entitled *Report of the Committee of Inquiry into Human Fertilisation and Embryology*, July 1984.
12. *Roe v. Wade*, 410 U.S. 113, 1973.
13. *Planned Parenthood of Southeastern Pa. v. Casey,* 505 U.S. 833, 1992.
14. *Whole Women's Health v. Hellerstedt*, 579 U.S. 274, 2016.
15. Pope: Ethical medical research requires morality, Safeguards Human Life, *CNA/EWTN News*, April 29, 2016; http://www.ncregister.com/site/print_article/49316/.
16. J. Robertson, Embryo screening for tissue matching, *Fertility and Sterility* 82, August, 2004: 290–291.
17. A. Sachedina, Islamic perspectives on the ethics of stem cell research, in D. L. Kleinman et al. (editors), *Controversies in Science & Technology: From Climate to Chromosomes*, Vol. 2, New Rochelle, NY: Mary Ann Liebert, Inc., 2008: 90–112.
18. C. B. Cohen, *Renewing the Stuff of Life: Stem Cells, Ethics, and Public Policy*, New York: Oxford University Press, 2007: 106–108.
19. Federal Policy for the Protection of Human Subjects. 45 CFR 46. 102–116, 1978.
20. HEW Support of Research Involving Human In Vitro Fertilization and Embryo Transfer: Report and Conclusions, Ethics Advisory Board, Department of Health, Education and Welfare, 1979; https://repository.library.georgetown.edu/handle/10822/811939.
21. J. C. Fletcher, U.S. public policy on embryo research: Two steps forward, one large step back, *Human Reproduction* 10, July 1995: 1875–1878.
22. Public Law 104-99, The Balanced Budget Downpayment Act, January 26, 1996, 110 Stat. 26.
23. W. J. Clinton, E.O. 12975, Protection of human research subjects and creation of national bioethics advisory commission, *Federal Registrar* 60, October 3, 1995: 52063.
24. The National Bioethics Advisory Commission (NBAC) Charter, 1996; https://repository.library.georgetown.edu/handle/10822/559325.

25. NBAC Reports, 1996–2001; https://repository.library.georgetown.edu/handle/10822/559325.
26. L. A. Meltzer, Human dignity and bioethics: Essays commissioned by the president's council on bioethics (Book Review), *New England Journal of Medicine* 359, August 7, 2008: 660–661.
27. B. Obama, E.O. 13521, Establishing the Presidential Commission for the Study of Bioethical Issues, *Federal Registrar* 74, November 24, 2009: 62671.
28. The President's Council on Bioethics; https://bioethicsarchive.georgetown.edu/pcbe/.
29. National Academy of Sciences and National Academy of Medicine, Committee on Human Gene Editing: Scientific, Medical, and Ethical Considerations; http://nationalacademies.org/gene-editing/consensus-study/.
30. A. Park, Life, the remix, *Time* 188, July 4, 2016: 42–48.
31. R. Benjamin, *People's Science: Bodies and Rights on the Stem Cell Frontier*, Stanford, CA: Stanford University Press, 2013.
32. M. Baker and P. Watson, Patients' organizations and their opinions: How much have they been taken into consideration when regulating stem cell research? in K. Hug and G. Hermeren (editors), *Translational Stem Cell Research: Issues beyond Debate on the Moral Status of the Human Embryo*, New York, NY: Springer Publishing, 2011: 365–373.
33. National Academy of Sciences, International summit on human gene editing: A global discussion, December, 2015; http://nationalacademies.org/gene-editing/Gene-Edit-Summit/index.htm.
34. Ibid.
35. Ibid.
36. J. Kimmelman et al., New ISSCR guidelines: Clinical translation of stem cell research, *Lancet* 387, May 14, 2016: 1979–1981; DOI: 10.1016/s0140-6736(16)30390-7.
37. National Academy of Sciences and National Academy of Medicine, Committee on Human Gene Editing: Scientific, Medical, and Ethical Considerations; http://nationalacademies.org/gene-editing/consensus-study/.
38. I. G. Cohen and E. Y. Adashi, Embryo disposition disputes: Controversies and case law, *Hastings Center Report* 465, 2016: 13–19.
39. B. Lo and L. Parham, Ethical issues in stem cell research, *Endocrine Review* 30, May, 2009: 204–213.
40. T. Caulfield et al., Research ethics and stem cells: Is it time to re-think current approaches to oversight? *EMBO Reports,* December 2014: 2–6; DOI: 10.15252/embr.201439819.
41. R. J. Blendon et al., The public, political parties, and stem-cell research, *New England Journal of Medicine* 365, November 17, 2011: 1853–1856.
42. M. E. Sleeboom-Faulkner, The large grey area between "bona fide" and "rogue" stem cell interventions—Ethical acceptability and the need to include local variability, *Technological Forecasting & Social Change* 109, 2016: 76–86.
43. The Presidential Commission for the Study of Bioethical Issues, *Bioethics for Every Generation: Deliberations and Education in Health, Science, and Technology*, Washington, DC: May 2016.
44. C. Funk et al., U.S. public wary of biomedical technologies to "Enhance" human abilities, *Pew Research Center*, July 26, 2016.

7 Conclusion

In our lifetime?

This book began by asking you, after educating yourself on the topic, to develop an informed opinion as to whether or not the isolation of human embryonic stem cells (hESCs) in 1998 would offer the opportunity for many with debilitating conditions to live longer and less painful lives. In order to answer this question, we have had to review cell biology, public opinion, moral concerns, funding, and policy, among other subjects.

Two privately funded U.S. research teams isolated hESCs in 1998. Science is all about *replication*, and like any other study, this new discovery had to be replicated to be validated. Soon other studies, both in the United States and abroad, would derive their own stem cell lines and begin the important task of transitioning from lab to clinic to cure.

In the U.S. the derivation was met with mixed public opinion. Scientists and individuals across all political spectra praised the study as a scientific first. James Thomson made the front cover of *Time* magazine, a publication read by millions. However, the initial enthusiasm was followed by moral concerns. After all, human embryos were being used.

In a recent study from July 2016, the Pew Research Center conducted a poll asking respondents about the potential use of three new biomedical technologies: gene editing, brain chip implants, and synthetic blood.[1] None of these are being used on humans. The study found that a majority of respondents were more "worried than enthusiastic" about all three technologies. The questions also asked the public about using these technologies to "enhance" human abilities. A different wording of the question might elicit a different response. For example, "Are you worried or enthusiastic about using one of these technologies to improve the health of a person who has a debilitating condition?" In Chapter 4, we analyzed how the wording of a question can change the context of the issue for the respondent, and affect his or her answer. None of the Pew questions asked specifically about hESC research, but the implications are evident. The general public can be suspicious about new and untested technologies, especially when they are linked with enhancing human abilities.

We discussed the distress of many when the first babies conceived via IVF were born. Louise Brown in the United Kingdom and Elizabeth Carr in the United States were considered freaks by some; 35 years later, Brown published her book and revealed that her life was quite normal aside from the criticism launched at her parents. Today, this once-controversial procedure has produced thousands of healthy children and beaming parents. It also led to a multimillion dollar industry that some think might be replicated when stem cell treatments become routine interventions.

But let us not forget the criticism heaped on the scientists who cloned Dolly, the famous or infamous sheep (depending on your point of view). Although Dolly died earlier than biologically produced sheep, a recently published study of her progeny confirms that they are "perfectly healthy as they turn 9."[2] Despite the fear that humans would be next, this has yet to happen. However, cloned animals are common in barns and racetracks around the world.

While praising the accomplishments of hESC researchers, President George W. Bush was concerned about the consequences of using humans for research. He shared his conflicted views with the American public. His address to the nation in 2001 is among the most cited examples of the art of compromise.[3] He was continually troubled by the research, but ultimately allowed studies to go forward to some extent.

The discovery of hESCs set off a veritable stampede of states and private companies clamoring to expand on Thomson's work. The discovery in 2007 of induced pluripotent stem cells (iPSCs) by Shinya Yamanaka was considered a great breakthrough and the perfect solution to the moral dilemma that still haunted the stem cell debate, because these hESC-like cells could be created by treating adult stem cells with a few transcription factors. Now scientists would no longer need to use human embryos to create these stem cell lines. Scientists were able to replicate Yamanaka's research and many expanded on his discovery by coaxing iPSCs to differentiate into specific tissue types; however, the value of hESCs remained undiminished among many in the scientific community; human embryonic stem cells are still referred to as the "gold standard" for scientific research on human development.[4] Some researchers questioned why both types of cells couldn't be used, prompting some to rely on iPSCs for their work.

The moral dilemma that faced President Bush is no longer a front-page story. Even at the height of the debate, polls revealed that more than half of the population believed that hESC funding was morally acceptable. With the exception of the Gallup polls, which continue to include a question about stem cells in their annual surveys, it is hard to find other polls that still ask about this topic although, as noted above, biomedical procedures are the topic of new survey research.

The concern that top U.S. scientists were fleeing the country to engage in stem cell research abroad also seems to have diminished. The evidence presented in Chapter 5 supports the contention that many stem cell scientists trained in other countries are coming to the United States to continue and expand their work. The International Society for Stem Cell Research (ISSCR) has facilitated a global sharing of information and talent with the hope that this international networking will lead to more rapid results.

In 2009, President Barack Obama issued an executive order expanding the number of hESC lines approved for federal funding. A review of the data in Chapter 5 reveals that today the National Institutes of Health (NIH) is spending more money on iPSC and hESC research. Scientists, depending on their focus, might start out using hESCs and then use iPSCs to replicate the same research.[5] Science is a curious pursuit that encourages exploring new ways to test hypotheses; we hope this book has instilled some of that curiosity in you.

Currently, the search for treatments continues. Human clinical trials using hESCs are few and limited in scope. There is only one trial using iPSCs taking place, and it is being undertaken in Japan by Shinya Yamanaka, the Nobel Prize-winning scientist who discovered iPSCs in 1998.[6] This trial was initiated and then suspended, but is now due to begin again in 2017. This trial focuses on age-related macular degeneration. This is also the focus of several of the hESC trials taking place. These trials are especially important given the increasingly large aging world population.

Today, the focus of the NIH and other research institutes, such as the California Institute for Regenerative Medicine (CIRM), is on translation. The strategic plans of both the NIH and CIRM emphasize the need to build partnerships with private sector companies. The focus is to move the research being done with hESCs and iPSCs from the labs into clinical trials and beyond. It should be reiterated that the most frequent trials using both types of cells do not involve people, but rather Petri dishes of specialized tissues derived from these cells. These trials are testing the safety and effectiveness of drugs for specific conditions and the effects these drugs have on other tissues in the body. In Chapter 5 we discussed these drug trials using stem cells. While these are not the cures that people were promised, one could hypothesize that using hESCs and iPSCs to improve the safety and effectiveness of existing and new drugs is a major contribution to medical science. Many expected this research to allow a person with a spinal cord injury to walk again; was this hype, or hope that science could produce results faster than was possible?

The hype has certainly been front and center. In order to convince the public and private sectors to fund their research, some scientists oversold their expected results, as did some of the journalists covering these discoveries. Everyone wants a positive finding, an uplifting story, or a miracle cure. Undoubtedly, some scientists really believed that within a few years of deriving hESCs, the key to treating Alzheimer's disease or spinal cord injuries could be found. After all, you can't spend your entire life looking for a cure and not have some shred of belief that it exists and will be discovered one day. You also read about the early years of hESC research, when just keeping the cells alive was a 24/7 endeavor that often required scientists to visit their labs in the middle of the night to replate their stem cell lines.

Their enthusiasm encouraged a cottage industry of clinics and labs around the world. Recently we learned that many of these "suspect" clinics are in the United States, pushing procedures under the rubric of "stem cell" treatments. This has fueled both the hope and the hype. This is not doublespeak: scientists have convinced us that they can, in time, improve our health, but there are charlatans who have convinced us that the treatments exist now. These unproved treatments administered in unlicensed clinics have cast a pall over the legitimate research. Some of these clinics are doing what they have always done, which is to provide cosmetic procedures, only now many of these procedures are framed under a different labeling or marketing terminology. However, some clinics are engaged in potentially dangerous procedures. The fact that both seem to be a booming business is evidence that many patients are filled with hope.

We are all waiting for the cure—that miracle treatment that will arrest Alzheimer's, stop Parkinson's, and restore mobility to thousands of afflicted people. Occasionally, a new interim discovery might make the science section of the evening news, but there are no miracle cures for the time being. Will they emerge? As President Obama noted, "maybe one day, maybe not in our lifetime, or even in our children's lifetime."[7] One day! Are you willing to wait?

Notes

1. C. Funk et al., U.S. public wary of biomedical technologies to "Enhance" human abilities, *Pew Research Center* July 26, 2016. Available at: http://www.pewinternet.org/2016/07/26/u-s-public-wary-of-biomedical-technologies-to-enhance-human-abilities/; G. Kolata, We don't trust scientists to make U.S. better, *New York Times* , July 27, 2016.
2. A recent study that examined the cloned daughters of Dolly the sheep concluded that cloned animals do not have shorter life spans; it seems that Dolly's early death was not due to cloning. K. D. Sinclair et al., Healthy aging of cloned sheep, *Nature Communications* 27, 7, July 26, 2016; J. Klein, Dolly's fellow clones, enjoying their golden years, *New York Times* A8, July 27, 2016.
3. E. Cohen, Bush's stem-cell ruling: A Missouri Compromise, *Los Angeles Times*, August 12, 2001. http://articles.latimes.com/2001/aug/12/opinion/op-33329; P. Cohen, Bush surprises with compromise on stem cells, *The New Scientist* August 10, 2001. https://www.newscientist.com/article/dn1142-bush-surprises-with-compromise-on-stem-cells/.
4. A. Azvolinsky, Not all stem cells created equal, *Scientist*, 1 28, July 2014; A. Gawrylewski, Embryonic stem cells still gold standard, *Scientist* 22, June 2008.
5. J. Yu et al., Induced pluripotent stem cell lines derived from human somatic cells, *Science* 318, December 21, 2007: 1917–1920.
6. J. Kyodo, Riken to resume retinal iPS transplant study in cooperation with Kyoto University, *The Japan Times* June 7, 2016. http://www.japantimes.co.jp/news/2016/06/07/national/science-health/riken-resume-retinal-ips-transplantation-cooperation-kyoto-univeristy/#.V5JUN4MrK70.
7. B. Obama, Remarks on signing an executive order removing barriers to responsible scientific research involving human stem cells and a memorandum on scientific integrity, March 9, 2009, *Public Papers of the Presidents of the United States*, Book 01, March 9, 2009: 199.

Glossary

This glossary has been *adapted* from the following sources: The National Institutes of Health, *Stem Cell Information* (www.stemcell.nih.gov/glossary); the International Society for Stem Cell Research, *Learn about Stem Cells* (www.isscr.org/visitor-types/public/stem-cell-glossary); and the European Union Consortium for Stem Cell Research, *EuroStemCell* (www.eurostemcell.org/stem-cell-glossary).

We have expanded the terms where appropriate with examples that link the terms to topics discussed in the text.

Adult stem cell Also called a somatic stem cell or tissue stem cell, this is an undifferentiated cell found in adult organisms such as bone marrow, skin, or intestine, with a limited capacity to (a) self-renew and (b) differentiate. An example: stem cells found in bone marrow can differentiate into white blood cells, red blood cells, and platelets and they can self-renew to produce more master cells that can also turn into white blood cells, red blood cells, and platelets.

Allogeneic transplant or cell therapy Transplanted cells or tissues that come from a single donor to treat a patient. The donor is usually a sibling or someone whose tissue type is a close match to that of the recipient. Most commonly used in bone marrow transplants. hESC therapies fall into this category. These therapies (used in animal models and in limited human clinical trials) are intended to become products that can use cells derived from a single donor to treat many unrelated patients.

Asymmetric division or asymmetric cell division The process that allows stem cells to self-renew (reproduce another stem cell), and to provide another cell that is specialized for the tissue in which it is located. This is a basic characteristic of stem cells.

Autologous transplant or cell therapy Transplanted cells or tissues that come from the patient's own body, involving the extraction of cells extra vivo (outside the body) and growing or manipulating them before transplanting them back into the patient. This procedure avoids problems associated with rejection as in allogeneic transplants. Using iPSC is an example of such a therapy.

Beta cells Are one of four major types of cells present in the pancreas of most mammals. Damage to these cells is the cause of Type I diabetes. This is an important research focus for stem cell researchers, in particular the Harvard Stem Cell Institute, where Douglas Melton is codirector and a top scientist studying stem cells and diabetes. The Harvard Stem Cell Institute and the private companies Viacyte, Inc. and Beta-02 Technologies are planning clinical trials for stem-cell-derived beta cells that are delivered via a capsule that protects the transplanted cells from the patient's immune system while they establish themselves and produce insulin.

Blastocyst An early embryo consisting of about 150 cells that has not yet implanted into the uterus. It is a spherical cell mass that contains a fluid-filled cavity, a cluster of cells called the inner cell mass (from which hESCs are derived), and is surrounded by an outer layer of cells that form the placenta. The blastocyst is totipotent, but hESC cells are pluripotent.

Cell culture The growth of tissue cells in artificial media. Growing or culturing stem cells has been difficult because knowledge about exactly what nutrients cells need was unknown. hESCs were initially grown over a layer of mouse fibroblasts bathed in calf serum. Today, stem cells are grown using human fibroblasts, the most common cells in connective tissue such as cartilage and filaments. The most serious problem in growing cells, especially those intended for transplantation into humans, continues to be contamination.

Chimeras Organisms that contain cells or tissues of other individuals of the same or different species. A common example is a mouse that has been injected with human cells so that it can be used for studying a human disease or testing a new drug. A person who receives a heart valve transplant from a pig is technically a chimera. The creation of chimeras for research has ethical implications. As long as human cells are injected into animals the concerns are reduced; the injection of animal cells into humans is more controversial.

Chromosome A threadlike strand of DNA that is encoded with genes that control how an organism grows and what it becomes. Human have 22 pairs of chromosomes plus the two sex chromosomes, XX in females and XY in males, for a total of 46.

Clinical trials A controlled test of a new drug, treatment, or therapy on human subjects that compare one drug, treatment, or therapy to another (after testing is done on animal models). Trials may or may not be randomized. The NIH maintains a list of all clinical trials both in the United States and abroad that can be accessed from their website: www.nih.clinicaltrials.gov.

Clone An identical copy of a cell (or organism), with the identical DNA sequences, that is produced when the nucleus of a cell is removed and replaced with the nucleus of a donor nucleus (see SCNT).

Cloning The process of producing genetically identical copies of a cell (therapeutic cloning), or a whole organism (reproductive cloning). In humans (or other mammals), nuclear transfer can be used to generate embryos with identical nuclear genetic material.

DNA Deoxyribonucleic Acid, a chemical found in the nucleus of cells that carries the instructions for making all the structures and materials the body needs to function: A long, chainlike molecule that transmits genetic information.

Ectoderm The outer layer of the three tissue layers of the early animal embryo that gives rise to the skin (epidermis), nervous system, tooth enamel, and lens of the eye.

Embryo The developmental stage between fertilization and the fetal stage. The embryonic stage ends 7–8 weeks after fertilization in humans.

Embryonic Stem Cell Research Oversight Committees (ESCRO Committees) In the 2005 Guidelines for Human Embryonic Stem Cell Research, the National Academy of Sciences called for the establishment of ESCRO Committees to review the ethical and legal issues concerning hESC research. This was designed to provide another layer of approval at research institutions working with hESCs, in addition to existing Institutional Review Boards (IRBs), that have been in existence since the 1970s to review and approve research involving human subjects.

Endoderm An embryonic tissue that gives rise to tissues such as the lungs, the intestine, the liver, and the pancreas.

Fetus Developmental stage from the end of the embryonic state, 7–8 weeks after fertilization, to the birth of the organism.

Fibroblast A common connective tissue that secretes the protein collagen and is found in most tissues of the body. Most cell cultures are derived from fibroblasts. These tissues are used as feeder cells that are able to support the growth of hESC because of the growth factors that they secrete. Early hESCs were grown on mouse fibroblasts, while more recent hESCs are derived using serums made from human fibroblasts.

Gastrulation An early phase in the development of an embryo in which cells proliferate and migrate to transform the inner cell mass of the blastocyst state into an embryo containing the three tissue layers—ectoderm, mesoderm, and endoderm.

Genome All of the genes that belong to a cell or an organism.

Germ cells Are pluripotent stem cells derived from the cells that give rise to eggs and sperm. John Gearheart derived his hESCs in 1998 from the germ line cells of an aborted fetus. Germ cells have properties similar to those of hESCs.

Germ layers These are the three layers, the endoderm, mesoderm, and ectoderm, that are formed in the embryo about 14 days after fertilization and give rise to all tissues of the body.

Graft-versus-host disease (GVHD) A condition that occurs after any allogeneic transplant in which the host's immune cells make antibodies against the donor's tissues. This is commonly associated with stem cell or bone marrow transplants but also applies to other forms of tissue grafts.

Human embryonic stem cells (hESC) Pluripotent cell derived from the inner cell mass of early stage human embryos (usually between day 4 or 5 after fertilization). When isolated from the embryo and grown in a lab, these cells can differentiate and divide indefinitely and give rise to all the tissues of all three germ layers.

Induced pluripotent stem cells (iPSC) Stem cells made by reprogramming adult cells (usually using skin or blood cells), back to their embryonic state using four transcription factors (genes), or by using their RNA equivalent. First developed by Shinya Yamanaka in 2005 using mouse skin cells, and in 2007 using human skin cells. These induced pluripotent stem cells (iPSCs) were created by the addition of four genes that "reprogrammed" an adult, differentiated cell back to an undifferentiated state of pluripotency.

Informed consent Requirement in the United States that patients give written consent before undergoing a medical procedure after being told about the risks and benefits. Individuals who donate excess embryos for research must be informed about the nature of the research, and that they will receive no financial benefits if commercial products are derived from their donated embryos. This presented a problem for some IVF clinics that had not obtained the necessary consent before giving embryos to research facilities.

Inner cell mass (ICM) The cluster of cells inside the blastocyst. These cells give rise to the embryo and ultimately the fetus. The ICM may be used to generate embryonic stem cells.

Institutional Review Board (IRB) The institutional committee at a university, government department, or private organization charged with reviewing research proposals to ensure that human participants are informed about their rights and protections.

In vitro Literally, in glass; experiments conducted in a lab. The growth of both hESCs and iPSCs in a dish or test tube is done in vitro.

In vitro fertilization (IVF) The procedure in which egg and sperm are brought together in a dish (in vitro). The fertilized egg (zygote) will start dividing, forming an embryo that can be implanted into the womb of a woman and give rise to a child. The excess embryos (after 4 or 5 days of division) are used to create hESCs.

Mesenchymal stem cells Multipotent adult stem cells that can differentiate into a variety of cell types including muscle, cartilage, fat, and bone. Bone marrow contains a number of progenitor cells known as mesenchymal stem cells that are capable of replication as undifferentiated cells or differentiated into bone, cartilage, fat, muscles, or tendons.

Mesoderm An embryonic tissue that gives rise to muscle, connective tissue, bones, blood, and many internal organs. The other two tissue layers are endoderm and ectoderm.

Multipotent Having the ability to develop into more than one cell type in the body. Bone marrow cells are multipotent because they can turn into red blood cells, white blood cells, or platelets.

Neural stem cells Are found in the brain and can make new nerve cells (neurons) and other cells that support nerve cells (glia). In an adult, neural stem cells are found in small areas of the brain where the replacement of nerve cells takes place. Scientists have only recently discovered that these neural cells can regenerate (albeit slowly) themselves. It was once thought that they were static and did not regenerate.

Niche "Stem cell niche" is a term used to describe a specific area of a tissue where the stem cell resides.

Parthenogenesis Reproduction, more common among insects, that occurs without the fusion of sperm and egg. Artificially inducing parthenogenesis with human eggs may be a means to isolate stem cells from an embryo without fertilization.

Passage or Passaging The transfer of cells that are being grown in a dish to another dish in order to continue growth. With hESCs this is done every 2–3 days, or the cells will begin to differentiate and lose their pluripotent quality. This is also called plating or replating, meaning that the cells are moved to another dish with fresh serum so they will continue to grow.

Plasticity The ability of stem cells from one type of adult tissue to generate other cells in an organism. This is a characteristic of both hESC and iPSC. Some adult stem cells can produce more than one cell type—blood stem cells, for example.

Pluripotent A stem cell that can differentiate into any of the three germ layers—endoderm, mesoderm, or ectoderm.

Preimplantation genetic diagnosis or testing (PGD) A procedure in which a cell is taken from an early embryo created in vitro and is tested for existing genetic diseases. The embryo can also be tested to determine a tissue match with another sibling. Chapter 2 discussed the Molly Nash case, in which PGD was used.

Primitive streak A thickening line that forms 14–15 days after fertilization and begins the process of differentiation into the three germ layers (ectoderm, mesoderm, and endoderm). The nervous system is the first structure to appear. This process was discussed by study commissions in England, Canada, and the United States as a basis for approving research on early embryos.

Progenitor cells An intermediate cell type between hESC and differentiated cells. Progenitor cells have the potential to give rise to a limited number or type of specialized cells and have reduced capacity for renewal. Blood and muscle cells are considered progenitor cells in building heart valves and blood vessels.

Randomized clinical trial A study in which the participants are assigned by chance to separate groups that compare different treatments; neither the researchers nor the participants can choose the group to which they will be assigned. Using chance to assign people in groups means that the groups will be similar and that the treatments they receive can be compared objectively. At the time of the trial, it is not known which treatment is best. In some trials, one treatment is a placebo (an inactive substance that looks the same and is administered in the same way as a drug in a clinical trial).

Regenerative medicine Multidisciplinary branch of medicine combining science, medicine, and engineering with the use of stem cells to replace, regenerate, or repair damaged tissues to restore normal function. Treatments can include cellular therapy, gene therapy, and tissue engineering.

Reproductive cloning Removing the nucleus of an egg cell and replacing it with the nucleus of an adult cell to create an embryo that is implanted into the uterus, where it becomes an organism identical to the donor. This has been done in animals. No human has been cloned (although a company, Clonaid, based in the Bahamas, claimed to have cloned a human in 2002. No proof was provided; U.S. scientists doubt it happened). Animal cloning does take place. In 2015 a Chinese Company, BoyaLife, located in Tianjin, announced it was building a plant scheduled to open in 2016 to clone animals for the commercial market.

Ribonucleic acid (RNA) A nucleic acid present in all living cells that is similar in structure to DNA. One of its main functions is to translate the genetic code of DNA into structural proteins.

Somatic cell nuclear transfer (SCNT) Also referred to as *nuclear transfer* or *cloning* takes place when the nucleus of an oocyte (egg cell) is removed and replaced with the nucleus of an adult (somatic) cell. An embryo is created that has the genome of the adult cell donor. If the embryo is allowed to develop to the blastocyst stage in vitro it can be used in both *therapeutic cloning* (a new hESC line can be developed that genetically matches the donor), or *reproductive cloning* (be implanted in the uterus of an animal to produce a genetic copy; Dolly the sheep was created using this method). Human reproductive cloning is legally prohibited in many countries, and ethically banned in most. Human therapeutic cloning is permitted depending on country or state.

Stem cells Cells with the ability to self-renew by cell division and giving rise to more cells like themselves (copies of themselves), and to divide and give rise to mature specialized cells that can make up every type of tissue and organ in the human or animal body. In 1961 an academic paper published by Drs. James Tillman and Ernest McCulloch was the first to prove the existence of stem cells by studying mouse models.

Teratocarcinoma A malignant tumor occurring most commonly in the ovaries or testes.

Teratoma A benign tumor with a mixture of cells from all three germ layers. These tumors may be spontaneous and are most often found in the ovary of a woman or testes of a man. In the lab, these tumors can be induced by injecting hESC into mice as evidence that the cells are pluripotent.

Therapeutic cloning Removing the nucleus of an egg cell and replacing it with the nucleus of an adult cell to create an embryo that is used to derive hESC. Once differentiated, these hESCs can be implanted into the donor to assist in the repair or replacement of tissues or organs. Although scientists report having cloned a human blastocyst, an early stage embryo, there was no attempt to create stem cell lines. At this time, we are not aware of any research in human trials.

Totipotent Stem cells that have the potential to develop into any cell found in the human body, including the placenta. The ability to form the placenta is a defining feature of totipotent cells. When an egg cell and a sperm cell merge they form a one-celled fertilized zygote that is capable of giving rise to all types of differentiated cells found in an organism, including the supporting extra-embryonic structures of the placenta. Four or five days after fertilization and several cycles of cell division, these totipotent cells begin to specialize. The inner cell mass, the source of embryonic stem cells, is pluripotent.

Transcription factor A general term referring to the wide assortment of proteins needed to initiate or regulate transcription.

Transcription The copying of a DNA sequence into RNA, catalyzed by the RNA polymerase.

Trophoblast The tissue of the developing embryo responsible for implantation and formation of the placenta. The trophoblast does not come from the inner cell mass, but from the cells surrounding it.

Umbilical cord blood stem cells Hematopoietic stem cells present in the blood of the umbilical cord during and shortly after delivery. Because these stem cells move from the liver, where blood-formation takes place during fetal life, to the bone marrow, where blood is made after birth, they can be used for the treatment of leukemia and other diseases of the blood. Private and public cord blood banks have emerged to collect these cells and store them for later use. However, there may not be enough umbilical cord stem cells in one sample to transplant into an adult. The first public umbilical cord blood bank was established in 1992 at the New York Blood Center with funding from the NIH and the National Heart, Lung, and Blood Institute. In 2005 Congress passed legislation to create a national inventory of cord blood samples. Today, there are hundreds of private cord blood banks where, for a fee, parents can store their child's cord blood indefinitely.

Unipotent Cells that can only turn into one type of cell, for example, skin cells, one of the most abundant type of cell in the body.

Zygote A one-celled embryo formed by the fusion of a sperm and egg. The zygote goes through several rounds of cell division before becoming an embryo (about 4–5 days in a human).

Bibliography

Many scientific articles are available from the National Center for Biotechnology Information (NCBI), National Library of Medicine (NLM), or National Institutes of Health (NIH). This database is funded by the U.S. Government to provide access to biomedical and genomic information. Sources that are available on the extensive database are designated with the url www.ncbi.nlm.nih.gov/ followed by information unique to the specific article.

Some scholarly articles are available by joining Research Gate at https://www.researchgate.net/ Joining is free. The site provides full text access to many peer-reviewed journals. Readers can also access many scholarly journal articles from a university or public library site.

Abate, T. 2001. UCSF stem cell expert leaving U.S./Scientist fears that political uncertainty threatens his research. *SFGate*, July 17. http://www.sfgate.com/health/article/UCSF-stem-cell-expert-leaving-U-S-Scientist-2899802.php.

Adelson, J. W. and J. K. Weinberg. 2010. The California stem cell initiative: Persuasion, politics, and public service. *American Journal of Public Health* 100(March): 446–451. http://www.ncbi.nlm.nih.gov/pmc/articles/PMC2820047/.

Administrative Procedures Act (APA). Pub. L. 79-404, 60 Stat. 237 June 11, 1946.

Alberta, H. B. et al. 2015. Assessing state stem cell programs in the United States: How has state funding affected publication trends? *Cell Stem Cell* 16 (2): 115–118. http://www.ncbi.nlm.nih.gov/pubmed/25658368.

Alberts, B. et al. 2013. *Essential Cell Biology*. New York, NY: Garland Science.

Alper, J. 2009. Geron gets green light for human trial of ES cell-derived product. *Nature Biotechnology* 27 (3): 213–214.

Alliance for Regenerative Medicine. http://www.alliancerm.org/member-profiles.

Altman, L. K. 1984. Test tube skin helps save 2 burn victims. *New York Times*, August 16. http://www.nytimes.com/1984/08/16/us/test-tube-skin-helps-save-2-burn-victims.html?pagewanted=all.

Alzheimer's Association. 2016. Alzheimer's Disease Facts and Figures. http://www.alz.org/documents_custom/2016-facts-and-figures.pdf.

American Academy of Arts and Sciences. 2014. *Restoring the Foundation: The Vital Role of Research in Preserving the American Dream*. Cambridge, MA. www.amacad.org/content/Research/research-project.aspx?d=1276.

American Academy of Pediatrics. 2007. Policy statement: Cord blood banking for potential future transplantation. *Pediatrics* 119 (1): 165–170.

American Catholic. 2001. Pope John Paul II addresses President Bush. News Feature. www.american-catholic.org/news/stemcell/pope-to-bush.asp.

American Society for Biochemistry and Molecular Biology. 2013. *Unlimited Potential: Vanishing Opportunity, Survey*. www.asbmb.org/uploadedFiles/Advocacy/Events/UPVO%20Report.pdf.

Asterias Biotherapeutics. 2016. Announces positive new long-term follow-up results for AST-OPC1. May 24. http://asteriasbiotherapeutics.com/.

Azvolinsky, A. 2014. Not all stem cells created equal. *Scientist* 128 (7). http://www.the-scientist.com/?articles.view/articleNo/40409/title/Not-All-Stem-Cells-Created-Equal/.

Baker, M. 2008. James Thomson: Shifts from embryonic stem cells to induced pluripotency. *Nature Reports Stem Cells*. http://www.nature.com/stemcells/2008/0808/080814/full/stemcells.2008.118.html.

Baker, M. and P. Watson. 2011. Patients' organizations and their opinions: How much have they been taken into consideration when regulating stem cell research? In K. Hug and G. Hermeren (editors), *Translational Stem Cell Research: Issues beyond Debate on the Moral Status of the Human Embryo*. New York: Springer Publishing: 365–373.

Baron, D. 2005. The global race for stem cell therapies. *BBC News*. world.org/technologystemcell/index.shtml.

Becker, A. J. et al. 1963. Cytological demonstration of the clonal nature of spleen colonies derived from transplanted mouse marrow cells. *Nature* 197 (4866): 452–454. http://www.ncbi.nlm.nih.gov/pubmed/24837151.

Bellomo, M. 2006. *The Stem Cell Divide: The Facts, the Fiction, and the Fear Driving the Greatest Scientific, Political, and Religious Debate of Our Time*. New York: AMACOM.

Benjamin, R. 2013. *People's Science: Bodies and Rights on the Stem Cell Frontier*. Stanford, CA: Stanford University Press.

Birmingham, K. 2003. Europe fragmented over embryonic stem cell research. *Journal of Clinical Investigation* 112 (4): 458. http://www.ncbi.nlm.hih.gov/pmc/articles/PMC171403.

Blendon, R. J. et al. 2011. The public, political parties, and stem-cell research. *New England Journal of Medicine* 365 (20): 1853–1856.

Blue Skies and the Future of Regenerative Medicine. 2015. February 1. https://www.youtube.com/watch?v=w5Yohe2jZd4.

Boadi, K. 2014. Erosion of funding for the National Institutes of Health threatens U.S. leadership in biomedical research. *Center for American Progress*, March 25. https://www.americanprogress.org/issues/economy/report/2014/03/25/86369/erosion-of-funding-for-the-national-institutes-of-health-threatens-u-s-leadership-in-biomedical-research/.

Bongso, A. et al. 1994. Fertilization and early embryology: Isolation and culture of inner cell mass cells from human blastocysts. *Human Reproduction* 9 (11): 2110–2117. http://humrep.oxfordjournals.org/content/9/11/2110.

Boseley, S. 2015. First UK patient receives stem cell treatment to cure loss of vision. *The Guardian*, September 29. https://www.theguardian.com/science/2015/sep/29/first-uk-patient-receives-stem-cell-treatment-to-cure-sight-loss.

Broad, W. J. 2004. U.S. is losing its dominance in the sciences. *New York Times, May 3*. http://www.nytimes.com/2004/05/03/us/us-is-losing-its-dominance-in-the-sciences.html?_r=0.

Brown, L. and M. Powell. 2015. *Louise Brown: My Life as the World's First Test-Tube Baby*. London: Tangent Books.

Bryant, J. 2002. BresaGen moves slowly to grow stem cell work. *Atlanta Business Chronicle*, November 11. http://www.bizjournals.com/atlanta/stories/2002/11/11/story3.html.

Bush, G. W. 2001. Address to the nation on stem cell research, August 9. In *Public Papers of the Presidents of the United States: George Bush, 2001*, Book II, July 1–December 31, 2001: 954. Washington, DC: Government Printing Office, 2001. https://georgewbush-whitehouse.archives.gov/news/releases/2001/08/20010811-1.html.

Bush, G. W. 2001. The president's radio address, August 11, 2001. In *Public Papers of the Presidents of the United States: George Bush, 2001*, Book II, July 1–December 31, 2001: 956–967. Washington, DC: Government Printing Office. https://georgewbush-whitehouse.archives.gov/news/releases/2001/08/text/20010811-1.html.

Bush, G. W. 2006. Message to the House of Representatives returning without approval to the Senate the Stem Cell Research Enhancement Act of 2005. In *Presidential Papers of the Presidents George W. Bush*, Book 02, Presidential Documents, July 1 to December 31, 2006, July 19: 1425. https://georgewbush-whitehouse.archives.gov/news/releases/2006/07/20060719-5.html.

Bush, G. W. 2007. E.O. 13435 of June 22, 2007, expanding approval of stem cell lines in ethically responsible ways. *Federal Register* 72 (120): 34591–34593. https://www.gpo.gov/fdsys/pkg/FR-2007-06-22/pdf/07-3112.pdf.

California Institute for Regenerative Medicine (CIRM). Clinical Trials. https://www.cirm.ca.gov/our-progress/funding-clinical-trials.

California Institute for Regenerative Medicine (CIRM). Derivation: RFA 12-03. https://www.cirm.ca.gov/our-funding/research-rfas/hipsc-derivation.

California Institute for Regenerative Medicine (CIRM). hESC Repository: RFA 12-04. https://www.cirm.ca.gov/our-funding/research-rfas/hpsc-repository.

California Institute for Regenerative Medicine (CIRM). Tissue Collection for Disease Modeling: RFA 12-02. https://www.cirm.ca.gov/our-funding/research-rfas/tissue-collection-disease-modeling.

California Institute for Regenerative Medicine (CIRM). Our Governing Board. https://www.cirm.ca.gov/board-and-meetings/board.

California Institute for Regenerative Medicine (CIRM). Search CIRM Grants. https://www.cirm.ca.gov/grants.

California Institute for Regenerative Medicine (CIRM). Stem Cell Agency Okays $150 Million "powerhouse." https://www.cirm.ca.gov/dailynewsclipslink/stem-cell-agency-okays-150-millionE2%80%98powerhouse%E2%80%99.

California Institute for Regenerative Medicine (CIRM). 2016. Beyond 2.0—Strategic Plan 2016 and Beyond. https://www.cirm.ca.gov/node/33649.

California Institute for Regenerative Medicine (CIRM). 2016. The Stem Cellar: The Official bBlog of CIRM, California's Stem Cell Agency. June 28. https://blog.cirm.ca.gov/2016/06/28/another-way-to-dial-back-stem-cell-hype-but-not-hope-put-a-dollar-figure-on-it/.

California Stem Cell Research & Cures Initiative Archives at University of California at Los Angeles. http://digital.library.ucla.edu/websites/2004_996_027/index.htm.

Cambridge Academic Content. Dictionary. www.dictionary.cambridge.org/us/dictionary/english/brain-drain.

Canadian Royal Commission on New Reproductive Technologies. 1994. Reproductive Technologies: Royal Commission, Final Report. April 22.

Canadian Royal Commission on New Reproductive Technologies. 1999. *Proceed with Care*. Ottawa.

Caulfield, T. et al. 2014. Research ethics and stem cells: Is it time to re-think current approaches to oversight? *EMBO Reports*, December. DOI: 10.15252/embr.201439819.

Caulfield, T. et al. 2016. Confronting stem cell hype: Against hyperbole, distortion, and overselling. *Science* 352 (628): 776–777. http://www.ncbi.nlm.nih.gov/pubmed/27174977.

CBS News. 2012. *Stem Cell Fraud*. January 12. https://www.youtube.com/watch?v=ovPZkQYee8Y.

CBS News Canada. 2015. Frustrated Former Patients Question Probes into Controversial Stem-Cell Researcher. http://www.cbc.ca/news/canada/manitoba/frustrated-former-patients-question-probes-into-controversial-stem-cell-researcher-1.3157549.

Cellular Dynamics International. 2013. Cellular Dynamics and Coriell Institute for Medical Research Awarded Multi-Million Dollar Grants from California Institute for Regenerative Medicine to Manufacture and Bank Stem Cell Lines. March 21. https://cellulardynamics.com/news-events/news-item/cellular-dynamics-and-coriell-institute-for-medical-research-awarded-multi-million-dollar-grants-from-california-institute-for-regenerative-medicine-to-manufacture-and-bank-stem-cell-lines/.

Chakma, J. et al. 2014. Asia's ascent—Global trend in biomedical R&D expenditures. *New England Journal of Medicine* 370 (1): 3–6. http://www.ncbi.nlm.nih.gov/pubmed/24382062.

Chin, M. H. et al. 2009. Induced Pluripotent stem cells and embryonic stem cells are distinguished by gene expression signatures. *Cell Stem Cell* 5 (1 and 2): 111–123.

Christopher and Dana Reeve Foundation. http://www.christopherreeve.org.

Clinton, W. J. 1994. Statement on federal funding of research on human embryos, December 2, 1994. In *Public Papers of the Presidents of the United States: William J. Clinton*, Book 11, August 1–December 31, 1994: 2459. https://www.gpo.gov/fdsys/pkg/WCPD-1994-12-12/pdf/WCPD-1994-12-12-Pg2459-3.pdf.

Clinton, W. J. 1995. E.O. 12975, Protection of human research subjects and creation of *National Bioethics Advisory Commission*. October 3. *Federal Register* 60 (193): 52063–52065. https://www.gpo.gov/fdsys/pkg/WCPD-1995-10-09/pdf/WCPD-1995-10-09-Pg1759.pdf.

Clymer, A. 2001. The nation: Wrong number; the unbearable lightness of public opinion polls. *New York Times*, July 22. http://www.nytimes.com/2001/07/22/weekinreview/the-nation-wrong-number-the-unbearable-lightness-of-public-opinion-polls.html.

Cohen, C. B. 2007. *Renewing the Stuff of Life: Stem Cells, Ethics, and Public Policy*. New York: Oxford University Press.

Cohen, E. 2001. Bush's stem-cell ruling: A Missouri Compromise. *Los Angeles Times*, August 12. http://articles.latimes.com/2001/aug/12/opinion/op-33329.

Cohen, I. G. and E. Y. Adashi. 2016. Embryo disposition disputes: Controversies and case law. *Hastings Center Report* 46 (5): 13–19.

Cohen, P. 2001. Bush surprises with compromise on stem cells. *The New Scientist*, August 10. https://www.newscientist.com/article/dn1142-bush-surprises-with-compromise-on-stem-cells/.

Colen, B. D. 2016. Potential diabetes treatment advances. *Harvard Gazette*, January 25. http://news.harvard.edu/gazette/story/2016/01/potential-diabetes-treatment-advances/.

Connolly, C. 2005. Frist breaks with Bush on stem cell research. *Washington Post*, July 30. http://www.washingtonpost.com/wp-dyn/content/article/2005/07/29/AR2005072900158.html.

Connolly, R. et al. 2014. Stem cell tourism—A web-based analysis of clinical services available to international travelers. *Travel Medicine and Infectious Disease* 12 (6 pt. B): 695–701. http://www.ncbi.nlm.nih.gov/pubmed/25449045.

Cook, M. 2015. Scepticism greets new stem-cell regulations in China. *BioEdge*, August 29. http://www.bioedge.org/bioethics/scepticism-greets-new-stem-cell-regulations-in-china/11547.

Cowan, C. H. et al. 2004. Derivation of embryonic stem-cell lines from human blastocysts. *New England Journal of Medicine* 350 (1): 1353–1356. http://www.ncbi.nlm.nih.gov/pubmed/14999088.

Cummings, B. J. et al. 2005. Human neural stem cells differentiate and promote locomotor recovery in spinal cord-injured mice. *Proceedings of the National Academy of Sciences* 102 (39): 14069–14074. http://www.ncbi.nlm.nih.gov/pubmed/16172374.

Cyranoski, D. 2013. Controversial stem-cell company moves treatment out of the United States. *Nature News*, 493, January 30. http://www.nature.com/news/controversial-stem-cell-company-moves-treatment-out-of-the-united-states-1.12332.

Cyranoski, D. 2014. Next-generation stem cells cleared for human trial. *Nature*, September 12. http://www.nature.com/news/next-generation-stem-cells-cleared-for-human-trial-1.15897.

Daley, G. Q. 2004. Missed opportunities in embryonic stem-cell research. *New England Journal of Medicine* 351: 627–628. http://www.ncbi.nlm.nih.gov/pubmed/15302910.

Daley, G. Q. 2004. Testimony by George Q. Daley, MD, PhD. Representing the American Society for Cell Biology before the U.S. Senate, Commerce Subcommittee on Science, Technology, and Space. September 29. http://www.ascb.org/wp-content/uploads/2015/10/GeorgeDaleytestimonybefore-SenateCommerceCommittee.pdf.

Davies, M. 2015. World's first "test-tube baby" reveals her mother received blood-splattered hate mail when she was born—Including a letter containing a plastic foetus. *Daily Mail*, July 24.

Deschler, B. and M. Lubbert. 2006. Acute myeloid leukemia: Epidemiology and etiology. *Cancer* 107 (9): 2099–2107. http://www.ncbi.nlm.nih.gov/pubmed/17019734.

Dreifus, C. 2006. At Harvard's stem cell center, the barriers run deep and wide. *New York Times*, January 24. http://www.nytimes.com/2006/01/24/science/at-harvards-stem-cell-center-the-barriers-run-deep-and-wide.html?_r=0.

Drew, D. 2005. Umbilical cord blood banking: A rich source of stem cells for transplant. *Advance for Nurse Practitioners* 13 (4): S2–S7.

Dye, T. R. 1990. *American Federalism: Competition among Governments*. Lexington, MA: Lexington Books.

Eckholm, E. 2014. No longer ignored, evidence solves rape cases years later. *New York Times*, August 2. http://www.nytimes.com/2014/08/03/us/victims-pressure-cities-to-test-old-rape-kits.html?_r=0.

Elias, P. 2004. America lags in stem cell research: Political pressure limits funds for cloning experiments. *NBC News*, updated February 13.

Entman, R. M. 1993. Framing: Toward clarification of a fractured paradigm. *Journal of Communication* 43 (4): 51–58.

Eperen, L. V., F. M. Marincola, and J. Strohm. 2010. Bridging the divide between science and journalism. *Journal of Translational Medicine* 8: 25–27. https://translational-medicine.biomedcentral.com/articles/10.1186/1479-5876-8-25.

Evans, M. J. and M. H. Kaufman. 1981. Establishment in culture of pluripotential cells from mouse embryos. *Nature* 292: 154–156.

Faison, A. M. 2005. The miracle of Molly. *Denver Magazine*, August. http://www.5280.com/magazine/2005/08/miracle-molly?page=full.

Federation of American Societies for Experimental Biology (FASEB). Federal Funding for Biomedical and Related Life Sciences Research. 2016. http://www.faseb.org/Portals/2/PDFs/opa/2016/Federal_Funding_Report_FY2017_FullReport.pdf.

Fiester, A. 2005. Ethical issues in animal cloning. *Perspectives in Biology and Medicine* 48 (2): 328–343.

Fletcher, J. C. 1996. U.S. public policy on embryo research: Two steps forward, one large step back. *Human Reproduction* 10: 1875–1878.

Fossett, J. W. 2007. Federalism by necessity: State and private support for human embryonic stem cell research. *Rockefeller Institute Policy Brief*, August 9. http://www.rockinst.org/pdf/health_care/2007-08-09-federalism_by_necessity_state_and_private_support_for_human_embryonic_stem_cell_research.pdf.

Fox, C. 2007. *Cell of Cells: The Global Race to Capture and Control the Stem Cell*. New York: Norton.

Fox, C. 2014. Embryonic stem cells in trial for diabetes. *Bioscience Technology*, October 16. http://www.biosciencetechnology.com/article/2014/10/embryonic-stem-cells-trial-diabetes.

Fremgen, B. F. 2016. *Medical Law and Ethics*, 5th ed. Boston: Pearson.

Friedman, D. 2014. Perk up: Facebook and Apple now pay for women to freeze eggs. *NBC News*, October 14. http://www.nbcnews.com/news/us-news/perk-facebook-apple-now-pay-women-freeze-eggs-n225011.

Fry-Revere, S. 2006. No tax money for stem cells. *Los Angeles Times*, November 28. www.latimes.com/news/la-oe-fry-revere28nov28-story.html.

Fry-Revere, S. and M. Elgin. 2008. Public stem cell research funding: Boon or boondoggle? Competitive Enterprise Institute: Issue Analysis 4, September 4: 1–24. https://cei.org/sites/default/files/Sigrid%20Fry-Revere%20and%20Molly%20Elgin%20-%20Public%20Stem%20Cell%20Research%20Funding.pdf.

Funk, C. et. al. 2016. U.S. Public wary of biomedical technologies to "Enhance" human abilities, Pew Research Center, July 26. http://www.pewinternet.org/2016/07/26/u-s-public-wary-of-biomedical-technologies-to-enhance-human-abilities/.

Furcht, L. and W. Hoffman. 2011. *The Stem Cell Dilemma: The Scientific Breakthroughs, Ethical Concerns, Political Tensions, and Hope Surrounding Stem Cell Research*. New York: Arcade Publishing.

Gallup. Stem cell research. http://www.gallup.com/home.aspx.

Gardner, A. 2010. Most Americans back embryonic stem cell research: Poll. *U.S. News and World Report*, October 7. http://health.usnews.com/health-news/managing-your-healthcare/research/articles/2010/10/07/most-americans-back-embryonic-stem-cell-research-poll.

Gavin, K. 2013. Five years after Michigan vote on human embryonic stem cells, U-M effort is in full swing. November 4. http://www.uofmhealth.org/news/archive/201311/stemcell5.

Gawrylewski, A. 2008. Embryonic stem cells still gold standard. *Scientist* 22 (6). http://www.the-scientist.com/?articles.view/articleno/26495/title/embryonic-stem-cells-still-gold-standard/.

Gearhart, J. 1998. New potential for human embryonic stem cells. *Science* 282 (5391): 1061–1062. http://www.ncbi.nlm.nih.gov/pubmed/9841453.

Gibbs, W. W. 2005. The California gambit. *Scientific American*, June 27. http://www.scientificamerican.com/article/the-california-gambit/.

Golden, F. and D. Thompson. 2001. Cellular biology: Stem winder. *Time* 158 (7): 26–29.

Gottweis, H. and R. Triedl. 2006. South Korean policy failure and the Hwang debacle. *Nature Biotechnology* 24 (2): 114–143.

Green, H. 2010. *Therapy with Cultured Cells*. Singapore: Pan Stanford Publishing.

Green, L. 2015. Embryo wars: Avoiding Sofia Vergara's legal woes. *MSNBC*, August 3. http://www.msnbc.com/msnbc/embryo-wars-sofia-vergara-legal-woes.

Green, M. R. and J. Sambrock. 2012. *Molecular Cloning: A Laboratory Manual*. Cold Spring Harbor, NY: Cold Spring Harbor Laboratory Press.

Green, R. M. 2007. *Babies by Design: The Ethics of Genetic Choice*. New Haven, CT: Yale University Press.

Green, R. M. 1995. Report of the human embryo research panel. *Kennedy Institute of Ethics Journal* 5 (1): 83–84. http://www.ncbi.nlm.nih.gov/pubmed/11645298.

Green, R. M. et al. 1996. The politics of human-embryo research. *New England Journal of Medicine* 335 (16): 1243–1244. https://repository.library.georgetown.edu/handle/10822/751498.

Gruen, L., L. Grabel, and P. Singer (editors). 2007. *Stem Cell Research: The Ethical Issues*. Malden, MA: Blackwell.

Gulbrandsen, C. E. et al. 2005. Legal framework pertaining to research creating or using human embryonic stem cells. In J. Odorico, S.-C. Zhang and R. Pedersen (editors), *Human Embryonic Stem Cells*, 332. New York: Garland Science/BIOS Scientific Publishers.

Gunter, K.C. et al. 2010. Cell therapy medical tourism: Time for action. *Cytotherapy* 12: 965–968.

Gurdon, J. B. 2006. From nuclear transfer to nuclear reprogramming: The reversal of cell differentiation. *Annual Review of Cell and Developmental Biology* 22 (1): 1–22.

Hamilton, A. 1787. The same subject continued: The insufficiency of the present confederation to preserve the union. Federalist No. 17. The Federalist Papers. The Library of Congress. http://thomas.loc.gov/home/histdox/fed_17.html.

Harvey, A. M. 1976. Johns Hopkins—The birthplace of tissue culture: The story of Ross G. Harrison, Warren Y. Lewis, and George O. Gey. *Johns Hopkins Medical Journal Supplement*: 114–123. http://www.ncbi.nlm.nih.gov/pubmed/801529.

Herold, E. 2006. *Stem Cell Wars: Inside Stories from the Frontlines*. New York: Palgrave Macmillan.

Herszenhorn, D. M. and C. Hulse. 2009. Deal struck on $789 billion stimulus plan. *New York Times*, February 12. http://www.nytimes.com/2009/02/12/us/politics/12stimulus.html?_r=0.

HEW support of research involving human in vitro fertilization and embryo transfer: Report and conclusion. https://repository.library.georgetown.edu/handle/10822/811939.

Hiltzik, M. 2012. Did the California stem cell program promise miracle cures? *Los Angeles Times*, May 30. http://articles.latimes.com/2012/may/30/news/la-mo-stem-cell-20120530.

Hiltzik, M. 2013. Stem cell agency not doing enough to avoid conflict of interest. *New York Times*, January 29. http://articles.latimes.com/2013/jan/29/business/la-fi-hiltzik-20120129.

Ho, S. S. et al. 2008. Effects of value predispositions, mass media use, and knowledge of public attitudes towards embryonic stem cell research. *International Journal of Public Opinion Research* 20 (2): 171–192.

Hopkins, P. D. 1998. Bad copies: How popular media represent cloning as an ethical problem. *Hastings Center Report* 28 (2): 6–13. http://www.ncbi.nlm.nih.gov/pubmed/9589288.

How Stuff Works. http://science.howstuffworks.com/life/genetic/cloning3.htm.

Hudson Institute of Medical Research. 2015. IVF pioneer, Alan Trounson, returns to the Institute. http://hudson.org.au/news-events/latest-news/ivf-pioneer-alan-trounson-returns-to-the-institute.

Hug, K. and G. Hermeren (editors.) 2011. *Translational Stem Cell Research: Issues beyond Debate on the Moral Status of the Human Embryo*. New York, NY: Springer Publishing.

Human, K. 2005. U.S. researchers recruited labs in China, South Korea, and England are attracting scientists frustrated by the lack of stem cell funding. *Denver Post*, May 25.

Hwang, W. S. et al. 2004. Evidence of a pluripotent human embryonic stem cell line derived from a cloned blastocyst. *Science* 303 (5664): 1669–1674.

Hwang, W. S. et al. 2005. Patient-specific embryonic stem cells derived from human SCNT blastocysts. *Science* 308 (5729): 1777–1783.

Illinois Regenerative Medicine Institute. http://www.stemcelldirectory.com/listing/Government/illinois-regenerative-medicine-institute-201.

International Council for Science. 2004. The Value of Basic Scientific Research. http://www.icsu.org/publications/icsu-position-statements/value-scientific-research/the-value-of-basic-scientific-research-dec-2004.

International Society for Stem Cell Research. 2008. *ISSCR Patient Handbook on Stem Cell Therapies*. www.isscr.org/publications/patient-handbook.

International Society for Stem Cell Research. 2016. Guideline for Stem Cell Research and Clinical Translation. May 12. http://www.isscr.org/docs/default-source/guidelines/isscr-guidelines-for-stem-cell-research-and-clinical-translation.pdf?sfvrsn=2.

Issa, A. M. 2014. 10 years of personalized medicine: How the incorporation of genomic information is changing practice and policy. *Personalized Medicine* 12 (1): 1–3.

Jackson, C. L. 2008. State pulls back on stem cell funding. *New Jersey Com.*, June 22. http://www.nj.com/newark/index.ssf/state_pulls_back_on_stem_cell.html.

Jacoby, J. 2010. Let the private sector fund stem-cell research. *Boston Globe*, August 29. http://archive.boston.com/bostonglobe/editorial_opinion/oped/articles/2010/08/29/let_the_private_sector_fund_stem_cell_research/.

James, S. D. 2013. Test tube baby Louise Brown turns 35. *ABC News*, July 25.

Jeffery, L. 2011. Human iPSC and ESC translation potential debated. *Nature Biotechnology* 29: 375–376. http://www.ncbi.nlm.nih.gov/pubmed/21552221.

Jensen, D. E. 2014. Evaluating California's stem cell experiment. *Sacramento Bee*, November 15. http://www.sacbee.com/opinion/california-forum/article3924977.html.

Johnson, C. Y. 2012. UMass stem cell lab to close. *Boston Globe*, June 28. http://archive.boston.com/business/healthcare/articles/2012/06/28/umass_stem_cell_bank_to_close_after_state_funding_runs_out/.

Johnson, J. A. and E. D. Williams. 2005. Stem cell research. *Congressional Research Service Report for Congress*, Updated August 10. http://fpc.state.gov/documents/organization/51131.pdf.

Johnson, J. A. and E. D. Williams. 2007. Stem cell research: Federal research funding and oversight. *Congressional Research Report for Congress*, RL 33540, April 18. https://www.fas.org/sgp/crs/misc/RL33540.pdf.

Jones, G. 2002. Bill boosting stem-cell research to be signed. *Los Angeles Times*, September 22. http://articles.latimes.com/2002/sep/22/local/me-stemcell22.

Jones, H. W., Jr. 2014. *In Vitro Fertilization Comes to America*. Williamsburg, VA: Jamestowne Bookworks.

Juan, S. 2015. Health authority announces step to rein in "wild" stem cell treatment. *China Daily*, August 21.

Kamel, R. M. 2013. Assisted reproductive technology after the birth of Louise Brown. *Journal of Reproduction and Infertility* 14 (3): 96–109.

Kamenova, K. and T. Caulfield. 2015. Stem cell hype: Media portrayal of therapy translation. *Science Translational Medicine* 7 (278): 278–228. https://www.researchgate.net/publication/273633864_Stem_cell_hype_Media_portrayal_of_therapy_translation.

Kaplan, K. 2016. Hundreds of companies in the U.S. are selling unproven stem cell treatments, study says. *Los Angeles Times*, June 30. http://www.latimes.com/science/sciencenow/la-sci-sn-unapproved-stem-cell-treatments-20160630-snap-story.html.

Kimmelman, J. et al. 2016. New ISSCR guidelines: Clinical translation of stem cell research. *Lancet* 387: 1979–1981. DOI: 10.1016/s0140-6736(16)30390-7.

Klein, J. 2016. Dolly's fellow clones, enjoying their golden years. *New York Times*, July 27.

Kleinman, D. L. et al. (editors). 2008. *Controversies in Science and Technology: From Climate to Chromosomes*, Vol. 2. New Rochelle, NY: Mary Ann Liebert, Inc.

Kline, R. M. 2001. Whose blood is it anyway? *Scientific American* 284 (4): 42–49. https://www.mcdb.ucla.edu/Research/Goldberg/HC70A_W03/pdf/Stem_Umbil.pdf.

Kneller, A. 2010. Stem cell treatment for burn patients earns Alpert Prize. *Harvard Medical School News*, September 3. http://hms.harvard.edu/news/stem-cell-treatment-burn-patients-earns-alpert-prize-9-3-10.

Knoepfler, P. 2013. Lessons from patients: Stem cell clinical trials unlikely options for most patients. *The Niche*, January 14. https://www.ipscell.com/2013/01/lessons-from-patients-stem-cell-clinical-trials-unlikely-option-for-most-patients/.

Knoepfler, P. 2015. How much do stem cell treatments really cost? https://www.ipscell.com/2015/02/stemcelltreatmentcost/.

Knoepfler, P. S. 2012. Key anticipated issues for clinical use of human induced pluripotent stem cells. *Regenerative Medicine* 7 (5): 713–720. http://www.ncbi.nlm.nih.gov/pubmed/22830621.

Knoepfler, P. S. 2013. *Stem Cells: An Insider's Guide*. Hackensack, NJ: World Scientific.

Knox, R. 2010. Offshore stem cell clinics sell hope, not science. *National Public Radio*, July 26. http://www.npr.org/2010/07/26/128696529/offshore-stem-cell-clinics-sell-hope-not-science.

Kolata, G. 2010. A cloning scandal rocks a pillar of science publishing. *New York Times*, July 7. http://www.nytimes.com/2005/12/18/world/asia/a-cloning-scandal-rocks-a-pillar-of-science-publishing.html.

Kolata, G. 2016. We don't trust scientists to make U.S. better. *New York Times*, July 27.

Kubota, C. et al. 2000. Six cloned calves produced from adult fibroblast cells after long-term culture. *Proceeding of the National Academy of Sciences* 97 (3): 990–995.

Kubota, C. et al. 2004. Serial bull cloning by somatic cell nuclear transfer. *Nature Biotechnology* 22 (1): 693–694.

Kyodo, J. 2016. Riken to resume retinal iPS transplant study in cooperation with Kyoto University. *The Japan Times*, June 7. http://www.japantimes.co.jp/news/2016/06/07/national/science-health/riken-resume-retinal-ips-transplantation-cooperation-kyoto-university/#.V5JUN4MrK70.

Lasky, L. 2001. Cord blood and our tomorrow. *American Association of Blood Banks: News*, March–April. http://www.aabb.org/Pages/default.aspx.

Lau, D. et al. 2008. Stem cell clinics online: The direct-to-consumer portrayal of stem cell medicine. *Cell Stem Cell* 3 (6): 591–594.

Lee, S. M. 2014. Stem cell agency acts to prevent potential conflict of interest. *SFGate*, July 9. https://www.sfgate.com/business/article/Stem-cell-agency-acts-to-prevent-potential-5610140.php.

Lefkowitz, J. P. 2008. Stem cells and the president—An inside account. *Commentary* 125 (1): 19–24. https://www.commentarymagazine.com/author/jay-lefkowitz/.

Len, S. 2004. South Korea, with renowned scientists, jolts field and revives debate. *New York Times*, February 13. http://www.nytimes.com/2004/02/13/us/cloning-stem-cells-laboratory-south-korea-with-renowned-scientists-jolts-field.html.

Levin, I. 1976. *The Boys from Brazil*. New York: Random House.

Levine, A. D. 2005. Trends in the geographic distribution of human embryonic stem-cell research. *Politics and the Life Sciences* 23 (2): 40–44.

Levine, A. D. 2011. Policy uncertainty and the conduct of stem cell research. *Cell Stem Cell* 8 (2): 132–135.

Levine, A. D. 2012. State stem cell policy and the geographic preferences of scientists in a contentious emerging field. *Science and Public Policy* 39 (4): 530–541.

Levine, A. D. and L. E. Wolf. 2012. The roles and responsibilities of physicians in patients' decisions about unproven stem cell therapies. *Journal of Law, Medicine & Ethics* 40 (1): 122–134.

Lewin, T. 2015. Egg donors challenge pay rates, saying they shortchange women. *New York Times*, October 17. http://www.nytimes.com/2015/10/17/us/egg-donors-challenge-pay-rates-saying-they-short change-women.html.

Lizotte, M.-K. 2015. The abortion attitudes paradox: Model specific and gender differences. *Journal of Women, Politics and Policy* 36 (1): 22–42.

Lo, B. and L. Parham. 2009. Ethical issues in stem cell research. *Endocrine Review* 30 (3): 204–213.

Loeb, N. 2015. Sofia Vergara's ex-fiancé: Our frozen embryos have a right to live. *New York Times*, April 30. http://www.nytimes.com/2015/04/30/opinion/sofiavergaras-ex-fiance-our-frozen-embryos-have-a-right-to-live.html.

Lujan, E. et al. 2015. Early reprogramming regulators identified by prospective isolation and mass cytometry. *Nature* 521 (14274): 352–356.

Marcotty, J. 2010. "Savior sibling" raises a decade of life-and-death questions. *Star Tribune*, September 22. http://www.startribune.com/savior-sibling-raises-a-decade-of-life-and-death-questions/103584799/.

Maron, D. F. 2016. Unproved stem cell clinics proliferate in the U.S. *Scientific American*, June 30. http://www.scientificamerican.com/article/unproved-stem-cell-clinics-proliferate-in-the-u-s/.

Marshall, E. 1998. A versatile cell line raises scientific hopes, legal questions. *Science* 282 (5391): 1014–1015.

Martin, G. R. 1981. Isolation of a pluripotent cell line from early mouse embryos cultured in medium conditioned by teratocarcinoma stem cells. *Proceedings of the National Academy of Sciences* 78 (12): 7643–7638.

Martin G. R. and M. J. Evans. 1975. Differentiation of clonal lines of teratocarcinoma cells: Formation of embryoid bodies in vitro. *Proceedings of the National Academy of Sciences* 72: 1441–1445.

Martin, M. 2002. Davis OKs stem cell research/California is first state to encourage studies. *San Francisco Chronicle, Sacramento Bureau*, September 23.

Maryland Technology Development Corporation. http://www.tedco.md.

Matthews, D. 2013. The sequester: Absolutely everything you could possibly need to know, in one FAQ. *Washington Post*, March 1. https://www.washingtonpost.com/news/wonk/wp/2013/02/20/the-sequester-absolutely-everything-you-could-possibly-need-to-know-in-one-faq/.

Maugh, T. H., II. 2009. Xiangzhong "Jerry" Yang dies at 49; leading researcher in cloning technology. *Los Angeles Times*, February 10. http://www.latimes.com/science/la-me-jerry-yang10-2009feb10-story.html.

McCulloch, E. A. and J. E. Till. 1960. The radiation sensitivity of normal mouse bone marrow cells, determined by quantitative marrow transplantation into irradiated mice. *Radiation Research* 13 (1): 115–125. http://citeseerx.ist.psu.edu/viewdoc/download?doi=10.1.1.470.6351&rep=rep1&type=pdf.

McDonald, J. 2009. Connecticut's continuing role in advocating stem cell research. *Yale Journal of Biology and Medicine* 82 (3): 97–99. http://www.ncbi.nlm.nih.gov/pmc/articles/PMC2744942/.

McGinley, L. 2016. Biden unveils launch of major, open-access database to advance cancer research. *Washington Post*, June 6. https://www.washingtonpost.com/national/health-science/biden-to-unveil-launch-of-major-open-access-database-to-advance-cancer-research/2016/06/05/8918c442-2b30-11e6-9de3-6e6e7a14000c_story.html.

McGinley, L. 2016. Unregulated stem-cell clinics proliferate across the U.S. *Washington Post*, July 1. https://www.washingtonpost.com/news/to-your-health/wp/2016/06/30/unregulated-stem-cell-clin ics-are-proliferating-across-the-u-s/.

McKinsey, J. 2016. Telephone Interview, NIH, Central Budget Office. May 12 and 13.

Meilaender, G. 2004. Stem cells and the Reagan legacy. *New Atlantis: A Journal of Technology and Society* Summer (6): 19–25. http://www.thenewatlantis.com/publications/stem-cells-and-the-reagan-legacy.

Melton, D. 2006. Understanding embryonic stem cells. Howard Hughes Medical Institute, 2006 Holiday Lecture, *Potent Biology: Stem Cells, Cloning and Regeneration*. https://www.hhmi.org/biointeractive/understanding-embryonic-stem-cells. Also at: https://www.youtube.com/watch?v=nYNBNZJ8Xck.

Meltzer, L. A. 2008. Human dignity and bioethics: Essays commissioned by the President's Council on Bioethics (Book review). *New England Journal of Medicine* 359 (6): 660–661.

Michigan Legislature, State Constitution, Section 27 Human Embryo and Embryonic Stem Cell Research. http://www.legislature.mi.gov/(S(qxhdcb5zkpoiotqsn0tsuill))/mileg.aspx?page=GetObject&objectname=mcl-Article-I-27.

Miller, C. C. 2014. Freezing eggs as part of employee benefits: Some women see darker message. *New York Times*, October 14. http://www.nytimes.com/2014/10/15/upshot/egg-freezing-as-a-work-benefit-some-women-see-darker-message.html.

Millman, J. 2014. Does it really cost $2.6 billion to develop a new drug? *Washington Post*, November 18. https://www.washingtonpost.com/news/wonk/wp/2014/11/18/does-it-really-cost-2-6-billion-to-develop-a-new-drug/.

Miniño, A. M. et al. 2002. *Deaths: Final Data for 2000*. National Vital Statistics Report, 50 (15). Hyattsville, MD: National Center for Health Statistics. https://www.cdc.gov/nchs/data/nvsr/nvsr50/nvsr50_15.pdf.

Morales, L. 2009. Majority of Americans likely support stem cell decision. *Gallup Report*. http://www.gallup.com/poll/116485/Majority-Americans-Likely-Support-Stem-Cell-Decision.aspx?g_source=stem%20cell%20research&g_medium=search&g_campaign=tiles.

Mummery, C. et al. 2001. *Stem Cells: Scientific Facts and Fiction*. London: Elsevier Inc.

Narioka, K. and P. Dvorak. 2013. Japan makes advances on stem-cell therapy. *Wall Street Journal*. http://www.wsj.com/articles/SB10001424127887323689204578571363010820642.

Nathan, V. 2015. Asterias's stem cell therapy shows promise in study. *Health*, August 31.

National Academy of Sciences. 2015. International Summit on Human Gene Editing: A Global Discussion. http://nationalacademies.org/gene-editing/Gene-Edit-Summit/index.htm.

National Academy of Sciences and National Academy of Medicine. 2016. Committee on Human Gene Editing: Scientific, Medical, and Ethical Considerations. http://nationalacademies.org/gene-editing/consensus-study/.

National Bioethics Advisory Commission. 1999. *Ethical Issues in Human Stem Cell Research*. Rockville, MD: U.S. Government Printing Office.

National Bioethics Advisory Commission (NBAC) Charter. 1996. https://repository.library.georgetown.edu/handle/10822/559325.

National Catholic Register. 2016. Pope: Ethical medical research requires morality, "Safeguards Human Life" *CNA/EWTN News*. http://www.ncregister.com/daily-news/pope-ethical-medical-research-requires-morality-safeguards-human-life.

National Conference of State Legislatures. 2015. http://www.ncls.org/research/elections-and-campaigns/chart-if-the-initiative-states.aspx.

National Conference of State Legislatures. 2016. Embryonic and Fetal Research Laws. http://www.ncsl.org/research/health/embryonic-and-fetal-research-laws.aspx.

National Institutes of Health. 2016. NIH-Wide Strategic Plan: Fiscal Years 2016–2020: Turning Discovery Into Health. https://www.nih.gov/sites/default/files/about-nih/strategic-plan-fy2016-2020-508.pdf.

National Institutes of Health, National Cancer Institute. 2016. National Cancer Act of 1971, Legislative History. http://www.cancer.gov/about-nci/legislative/history.

National Institutes of Health, National Human Genome Research Institute. 2016. Fact Sheets on Science, Ethics and the Institute. https://www.genome.gov/10000202.

National Institutes of Health Reform Act of 2006, P.L 109-482. January 15, 2007; 120 Stat. 3675. https://www.nih.gov/about-nih/who-we-are/nih-reauthorization.

National Science Foundation. 2014. Report to the National Science Board on the National Science Foundation's Merit Review Process. Fiscal Year 2013. http://www.nsf.gov/pubs/2014/nsb1432/nsb1432.pdf.

Nelkin, D. 1995. *Selling Science: How the Press Covers Science and Technology*. New York: W.H. Freeman and Company.

Newport, F. 2009. Catholics similar to mainstream on abortion, stem cells. *Gallup Polls*. http://www.gallup.com/poll/117154/Catholics-Similar-Mainstream-Abortion-Stem-Cells.aspx?g_source=stem%20cells&g_medium=search&g_campaign=tiles.

New State Ice Company v. Liebmann. 285 U.S. 262, 1932.

New York, Department of Health, New York State Stem Cell Science. 2016. http://www.stemcell.ny.gov/about-us-O).

New York Stem Cell Foundation. 2016. http://www.nyscf.org.

Nightlight Christian Adoptions. https://www.nightlight.org.

NIH. 1994. Report of the Human Embryo Research Panel, Vol. I. https://repository.library.georgetown.edu/bitstream/handle/10822/559352/human_embryo_vol_1.pdf?sequence=1&isAllowed=y.

NIH. 1999. National Bioethics Advisory Commission, Ethical Issues in Human Stem Cell Research—Volume 1—Report and Recommendations of the NBAC Volume 2; Commissioned Papers (January 2000); Volume 3; Religious Perspectives (June 2000). http://www.hhs.gov/asl/testify/t970508c.html.

NIH. 2000. National Institutes of Health guidelines for research using human pluripotent stem cells. *Federal Register*, 65(166): 51975. http://stemcells.nih.gov/staticresources/news/newsArchives/fr25au00-136.htm.

NIH. 2001. Notice of Criteria for Federal funding of Research on Existing Human Embryonic Stem Cells and Establishment of NIH Human Embryonic Stem Cell Registry. November 7 NOT-OD-02-005, Office of the Director, NIH. https://grants.nih.gov/grants/guide/notice-files/NOT-OD-02-005.html.

NIH. 2001. Plan for Implementation of E.O. 13435. September 18, 2007. http://stemcells.nih.gov/staticresources/policy/eo13435.pdf.

NIH. 2001. Stem Cell and Scientific Progress and Future Research Directions. June 2001. stemcells.nih.gov/staticresources/info/scirport/pdfs/fullrptstem/pdf.

NIH. 2001. Update on Existing Human Embryonic Stem Cells (August 27, 2001); NIH, Federal Government Clearances for Receipt of International Shipment of Human Embryonic Stem Cells. November 16, 2001, NOT-OD-02-013. http://stemcells.nih.gov/policy/statements/pages/082701list.aspx.

NIH. 2003. NIGMS Center Grants to Explore Stem Cell Biology. September 29, 2003. https://www.nigms.nih.gov/News/results/Pages/HESC.aspx.

NIH. 2007. Human Embryonic Stem Cell Registry. http://grants.nih.gov/stem_cells/registry/current.htm.

NIH. 2008. Office of legislative policy analysis legislative updates. Pending Legislation—110th Congress. https://olpa.od.nih.gov/legislation/110/pendinglegislation/stem_cell_research.asp.

NIH. 2009. Guidelines on human stem cell research. In stem cell information. *Federal Register* 74 (128): 32170–32175. http://stemcells.nih.gov/policy/pages/2009guidelines.aspx.

NIH. 2015. Clinical Research Trials and You. November 16, 2015. www.nih.gov/health-information/nih-clinical-research-trials-you/basics.

NIH. 2016. A Short History of the National Institutes of Health. http://history.nih.gov/exhibits/history/.

NIH. 2016. About NIH, What we do, Budget. www.nih.gov/about-nih/what-we-do/budget.

NIH. 2016. Clinical Centers. http://clinicalcenters.nih.gov.

NIH. 2016. Finding a Clinical Trial. https://www.nih.ov/health-information/nih-clinical-trials-you/finding-clinical-trial.

NIH. 2016. National Cancer Institute. www.cancer.gov.

NIH. 2016. National Center for Advancing Translational Sciences. http://ncats.nih.gov.

NIH. 2016. National Eye Institute, Statistics and Data. https://nei.nih.gov.

NIH. 2016. National Institute of General Medical Sciences. www.nigms.nih.gov/pages/default.aspx.#.

NIH. 2016. NIH. National Library of Medicine, Medical Encyclopedia. https://medlineplus.gov/ency/article/001309.htm.

NIH. 2016. NIH-Wide Strategic Plan: Fiscal Years 2016–2020. https://www.nih.gov/sites/default/files/about-nih/strategic-plan-fy2016-2020-508.pdf.

NIH. 2016. Research Portfolio Online Reporting Tools (RePort). https://projectreporter.nih.gov/reporter_searchresults.cfm.

NIH. 2016. State Initiatives for Stem Cell Research. http://stemcells.nih.gov/research/pages/stateResearch.aspx.

NIH. 2016. Stem Cell Information, Stem Cell Basic. www.stemcells.nih.gov/info/basics/pages/basics3.aspx.

NIH. 2016. The Human Embryonic Stem cell and the Human Embryonic Germ Cell. http://stemcells.nih.gov/info/scireport/pages/chapter3.aspx.

NIH, National Library of Medicine. https://medlineplus.gov/ency/imagepages/8682.htm.

Nisbet, M. C. et al. 2003. Framing science: The stem cell controversy in an age of press/politics. *The Harvard International Journal of Press/Politics* 8: 36–70. https://www.researchgate.net/publication/249809264_Framing_ScienceThe_Stem_Cell_Controversy.

Nisbet, M. C. 2004. The polls—Trends: Public opinion about stem cell research and human cloning. *Public Opinion Quarterly* 68 (1): 132–155. https://www.researchgate.net/publication/237267137_Public_Opinion_About_Stem_Cell_Research_and_Human_Cloning.

Nisbet, M. C. 2004. Understanding what the American public really thinks about stem cell and cloning research. *Science and the Media*, The Committee for Skeptical Inquiry. http://www.csicop.org/specialarticles/show/understanding_what_the_american_public_really_thinks.

Nisbet, M. C. 2005. The competition for worldviews: Values, information, and public support for stem cell research. *International Journal of Public Opinion Research* 17 (1): 90–112. https://www.researchgate.net/publication/31015667_The_Competition_for_Worldviews_Values_Information_and_Public_Support_for_Stem_Cell_Research.

Nisbet, M. C. and A. B. Becker. 2014. The polls –Trends: Public opinion about stem cell research, 2002 to 2010. *Public Opinion Quarterly* 78 (4): 1003–1022.

Nisbet, M. C. and E. M. Markowitz. 2014. Understanding public opinion in debates over biomedical research: Looking beyond political partisanship to focus on beliefs about science and society. *PLoS One* 9 (2). http://www.ncbi.nlm.nih.gov/pmc/articles/PMC3928253/.

Nisbet, M. C. and C. Mooney. 2007. Framing science. *Science* 316 (5821): 56. http://science.sciencemag.org/content/316/5821/56.

Nisbet, M. C. and C. Mooney. 2007. Thanks for the facts: Now sell them. *Washington Post*, April 15. http://www.washingtonpost.com/wp-dyn/content/article/2007/04/13/AR2007041302064.html.

Obama, B. 2009. E.O. 13505 of March 9, 2009, Removing barriers to responsible scientific research involving human stem cells. *Federal Register* 74 (46): 10667–10668. https://www.gpo.gov/fdsys/pkg/FR-2009-03-11/pdf/E9-5441.pdf.

Obama, B. 2009. Remarks of the president—As prepared for delivery- signing of stem cell executive order and scientific integrity presidential memorandum. https://www.whitehouse.gov/the-press-office/remarks-president-prepared-delivery-signing-stem-cell-executive-order-and-scientifi.

O'Brien, N. F. 2001. Embryonic stem-cell research immoral, unnecessary, bishops say. *Catholic News Service*, June 13. www.americancatholic.org/news/stemcell/#background.

Obokata, H. et al. 2014. Bidirectional development potential in reprogrammed cells with acquired pluripotency. *Nature* 505 (7485): 676–680. http://www.nature.com/nature/journal/v505/n7485/full/nature12969.html.

Obokata, H. et al. 2014. Stimulus-triggered fate conversion of somatic cells into pluripotency. *Nature* 505 (7485): 641–647. http://www.nature.com/nature/journal/v505/n7485/full/nature12968.html.

Pagliuca, F. W. et al. 2014. Generation of functional human pancreatic B cells in vitro. *Cell* 159 (2): 428–439. http://www.cell.com/abstract/S0092-8674(14)01228-8.

Palca, J. 2007. States take lead in funding stem-cell research. *National Public Radio, March 30.* http://www.npr.org/templates/story/story.php?storyId=9244363.

Park, A. 2005. Dogged pursuit. *Time*, November 10: 217.

Park, A. 2011. *The Stem Cell Hope: How Stem Cell Medicine Can Change Our Lives.* New York: Hudson Street Books.

Park, A. 2016. Life, the remix. *Time*, July 4: 42–48.

Parson, A. B. 2004. *The Proteus Effect: Stem Cells and Their Promise for Medicine.* Washington, DC: Joseph Henry Press.

Patton, K. T. and G. A. Thibodeau. 2013. *The Human Body in Health and Disease*. Atlanta, GA: Elsevier.

Pew Research Center. 2002. Public Makes a Distinction on Genetic Research: Cloning Opposed, Stem Cell Research Narrowly Supported. http://www.people-press.org/files/legacy-pdf/152.pdf.

Pew Research Center. 2005. Strong Support for Stem Cell Research; Abortion and Rights of Terror Suspects Top Court Issues. http:/www.people-press.org/files/legacy-pdf/253.pdf.

Pew Research Center. 2016. U.S. Public Wary of Technologies to "Enhance" Human Abilities. http:/www.pewresearch.org.

Pfizer, Centers for Therapeutic Innovation. An Introduction to the Centers sums up the goal of the Centers as—Translating Leading Science into Clinical Candidates Through Networked Collaboration. http://www.pfizer.com/research/rd_partnering/centers_for_therapeutic_innovation.

Phillips, R. et al. 2013. *Physical Biology of the Cell.* New York: Garland Science Publishing.

Picoult, J. 2003. *My Sister's Keeper*. New York: Atria.

Planned Parenthood of Southeastern *Pa. v. Casey*. 505 U.S.833, 846, 1992.

Pollack, A. 2006. Trial over California stem cell research ends. *New York Times*, February 27. www.nytimes.com/2006/03/03/us/trial-over-caifornia-stem-cellresearch-ends.html.

Pollock, A. 2007. California stem cell research is upheld by appeals court. *New York Times* March 3. www.nytimes.com/2007/02/27/us/27stem.html.

Pope Paul VI. 1968. Humanae Vitae. On the Regulation of Birth. http://www.papalencyclicals.net/Paul06/p6humana.htm.

Prakash, S. and V. Valentine. 2007. Timeline: The rise and fall of Vioxx. *National Public Radio*, November 10. http://www.npr.org/templates/story/story.php?storyId=5470430.

Presidential Commission for the Study of Bioethical Issues. 2016. *Bioethics for Every Generation: Deliberations and Education in Health, Science, and Technology*. Washington, DC: Presidential Commission for the Study of Bioethical Issues.

President's Council on Bioethics. 2002. *Human Cloning and Human Dignity: An Ethical Inquiry*. Washington, DC: President's Council on Bioethics.

Priest, S. H. and T. Ten Eyck. 2003. News coverage of biotechnology debates. *Society* 40 (6): 29–34, 29.

Pub. L. 103-43, June 10, 1993. NIH Revitalization Act of 1993. https://www.govtrack.us/congress/bills/103/s1.

Pub. L. 104-99 128, Stat. 26, 34, 1996. The Balanced Budget Downpayment Act, I. https://www.congress.gov/104/plaws/publ99/PLAW-104publ99.pdf.

Pub. L. 105-119 617, November 26, 1997. 111 Stat. 2519. Departments of Commerce, Justice, and State, the Judiciary, and Related Agencies Appropriations Act. 1998. https://www.govtrack.us/congress/bills/105/hr2267.

Pub. L. 109-129 119 Stat. 2552, December 20, 2005. Stem Cell Therapeutic and Research Act of 2005. https://www.congress.gov/bill/109th-congress/house-bill/2520.

Pub. L. 109-242, 120 Stat. 570-571, July 20, 2006. Fetus Farming Prohibition Act of 2006. https://www.congress.gov/109/plaws/publ242/PLAW-109publ242.pdf.

Qiang, L. et al. 2014. Instant neurons: Directed somatic cell reprogramming models of central nervous system disorders. *Biological Psychiatry* 75 (12): 945–951.

Rabb, H. 1999. Letter from U.S. Department of Health and Human Services, Office of the General Counsel, Harriet Rabb to Harold Varmus, Director, National Institutes of Health, January 15. profiles.nlm.nih.gov/ps/retrieve/Narrative/MV/p-nid/191/p-docs/true.

Rasko, J. and C. Power. 2015. What pushes scientists to lie? The disturbing but familiar story of Haruko Obokata. *The Guardian*. https://www.theguardian.com/science/2015/feb/18/haruko-obokata-stap-cells-controversy-scientists-lie.

Reis, R. 2008. How Brazilian and North American Newspapers frame the stem cell research debate. *Science Communications* 29 (3): 316–334. https://www.researchgate.net/publication/242298040_How_Brazilian_and_North_American_Newspapers_Frame_the_Stem_Cell_Research_Debate.

Research!America. Polls and Publications. www.researchamerica.org/polls-and-publications.

Rheinwald, J. G. and H. Green. 1975. Serial cultivation of strains of human epidermal keratinocytes: The formation of keratinizing colonies from single cells. *Cell*, November: 331–343.

Riordan, D. G. 2008. Perspective: Research funding via direct democracy: Is it good for science? *Issues in Science and Technology* 24 (4). www.issues.org/24-4/p_riordan/.

Robertson, J. 2004. Embryo screening for tissue matching. *Fertility and Sterility* 82 (2): 290–291. https://law.utexas.edu/faculty/jrobertson/Robertson6.pdf.

Rockey, S. 2013. Application success rates decline in 2013. *National Institutes of Health Office of Extramural Research Blog*, December 18. http://nexus.od.nih.gov/all/2013/12/18/application-success-rates-declinein-2013/.

Roe v. Wade. 410 U.S. 113. 1973.

Roman Reed Foundation: Hope for Spinal Cord Research. http://www.romanreedfoundation.com.

Roush, W. 2005. Genetic savings and clone: No pet project. *MIT Technology Review* 1,085: 31.

Ruse, M. and C. A. Pynes (editors). 2006. *The Stem Cell Controversy: Debating the Issues*, 2nd ed. Amherst, NY: Prometheus Books.

Rutgers, School of Arts and Sciences, W.M. Keck Center for Collaborative Neuroscience. The Spinal Cord Injury Project. http://keck.rutgers.edu/research-clinical-trials/research-clinical-trials.

Ryan, K. J. 2006. The Politics and Ethics of Human Embryo and Stem Cell Research. In M. Ruse and C. A. Pynes (editors), *The Stem Cell Controversy*, 2nd ed. Amherst, NY: Prometheus Books: 291–300.

Saad, L. 2005. Americans OK with using embryos in medical research. *Gallup Polls*. http://www.gallup.com/poll/16486/Americans-Using-Embryos-Medical-Research.aspx?g_source=stem%20cells&g_medium=search&g_campaign=tiles.

Sachedina, A. 2008. Islamic perspectives on the ethics of stem cell research. In D. L. Kleinman et al. (editors), *Controversies in Science and Technology: From Climate to Chromosomes*, Vol. 2. New Rochelle, NY: Mary Ann Liebert, Inc.: 90–112.

Santora, M. 2003. Researchers on stem cells are making do, and hoping. *New York Times*, September 17.

Scheufele, D. A. 1999. Framing as a theory of media effects. *Journal of Communication* 49 (1): 103–122. https://www.researchgate.net/publication/209409815_Framing_As_a_Theory_of_Media_Effects.

Schmeck, H. M., Jr. 1988. DNA and crime: Identification from a single hair. *New York Times*, April 12. http://www.nytimes.com/1988/04/12/science/dna-and-crime-%09identification-from-a-single-hair.html.

Schwartz, J. and A. Devroy. 1994. Clinton to ban U.S. funds for some embryo studies. *Washington Post*, December 3. http://www.washingtonpost.com/wp-dyn/content/article/2005/07/27/AR2005072701318.html.

Schwartz, S. D. et al. 2012. Embryonic stem cell trials for macular degeneration: A preliminary report. *Lancet* 379 (9817): 713–720. http://www.ncbi.nlm.nih.gov/pubmed/22281388.

Shamblott, M. J. et al. 1998. Derivation of pluripotent stem cells from cultured human primordial germ cells. *Proceedings of the National Academy of Sciences* 95 (23): 13726–13731. http://www.ncbi.nlm.nih.gov/pubmed/9811868.

Shan, J. 2015. Health authority announces step to rein in "wild" stem cell treatment. *China Daily*, August 21.

Shapiro, B. 2015. *How to Clone a Mammoth: The Science of De-Extinction*. Princeton, NJ: Princeton University Press.

Sheldon, S. and S. Wilkinson. 2004. Hashmi and Whitaker: An unjustifiable and misguided distinction? *Medical Law Review* 12 (2): 137–163. http://medlaw.oxfordjournals.org/content/12/2/137.citation.

Sheldon, S. and S. Wilkinson. 2014. Should selecting savior siblings be banned? *Journal of Medical Ethics* 30 (6): 533–537.

Sherley v. Sebelius, 610 F.3d 69, 73 (D.C. Cir. 2010). https://www.cadc.uscourts.gov/internet/opinions.nsf/6c690438a9b43dd685257a64004ebf99/$file/11-5241-1391178.pdf.

Simon, S. 2016. Annual Report to the Nation: Cancer Death Rates Continue to Decline; *Increase in Liver Cancer Deaths Cause For Concern. American Cancer Society*. http://www.cancer.org/latest-news/cancer-statistics-report-death-rate-down-23-percent-in-21-years.html.

Sinclair, K. D. et al. 2016. Healthy ageing of cloned sheep. *Nature Communications* 27 (7). http://www.nature.com/articles/ncomms12359.

Skloot, R. 2010. *The Immortal Life of Henrietta Lacks*. New York: Random House.

Slack, J. 2012. *Stem Cells: A Very Short Introduction*. New York: Oxford University Press.

Sleeboom-Faulkner, M. E. 2016. The large grey area between "bona fide" and "rogue" stem cell interventions—Ethical acceptability and the need to include local variability. *Technological Forecasting & Social Change* 109: 76–86.

Smith, K. P. et al. 2009. Pluripotency: Toward a gold standard for human ES and iPS cells. *Journal of Cellular Physiology* 220 (1): 21–29. http://www.ncbi.nih.gov/pubmed/19326392.

Solomon, S. 2012. Realizing the Promise of Stem Cell Research. http://nyscf.org/susansolomontedtalk.

Somers, T. 2007. Defeat in N.J. of stem cell initiative raises alarm. *San Diego Union Tribune*, November 11. http://www.sandiegouniontribune.com/uniontrib/20071111/news_1b11njstems.html.

Spar, D. and A. Harrington. 2007. Selling stem cell science: How markets drive law along the technological frontier. *American Journal of Law & Medicine* 33 (4): 541–565.

Stengel, R. 2008. The Time 100 team. *Time*, May 12: 96.

Stewart, A. M. and G. W. Kneale. 2000. A-bomb survivors: Factors that may lead to a reassessment of radiation hazard. *International Journal of Epidemiology* 20 (4): 708–714. http://www.ncbi.nlm.nih.gov/pubmed/10922349.

Stewart, C. O. et al. 2009. Beliefs about science and news frames in audience evaluations of embryonic and adult stem cell research. *Science Communication* 30 (4): 427–452. https://www.researchgate.net/publication/249678864_Beliefs_About_Science_and_News_Frames_in_Audience_Evaluations_of_Embryonic_and_Adult_Stem_Cell_Research.

Stewart, C. O. 2013. The influence of news frames and science background on attributions about embryonic and adult stem cell research: Frames as heuristic/biasing cues. *Science Communication* 35 (1):

86–114. https://www.researchgate.net/publication/258186584_The_Influence_of_News_Frames_and_Science_Background_on_Attributions_About_Embryonic_and_Adult_Stem_Cell_Research_Frames_as_Heuristic_Biasing_Cues.

Sullivan, G. M. and A. R. Artino, Jr. 2013. Analyzing and interpreting data from Likert-type scales. *Journal of Graduate Medical Education* 5 (4): 541–542.

Takahashi, K. and S. Yamanaka. 2006. Induction of pluripotent stem cells from mouse embryonic and adult fibroblast cultures by defined factors. *Cell* 126 (4): 663–676. http://www.ncbi.nlm.nih.gov/pubmed/16904174.

The Michael J. Fox Foundation for Parkinson's Research. http://www.michaeljfox.org.

Thomson, J. A. et al. 1998. Embryonic stem cell lines derived from human blastocysts. *Science* 282 (5391): 1145–1147. http://www.ncbi.nlm.nih.gov/pubmed/9804556.

Till, J. E. and E. A. McCulloch. 1961. A direct measurement of the radiation sensitivity of normal mouse bone marrow cells. *Radiation Research* 14 (2): 213–222. http://www.ncbi.nlm.nih.gov/pubmed/13776896.

Todaro, G. J. and H. Green. 1963. Qualitative studies of the growth of mouse embryo cells in culture and their development into established lines. *Journal of Cell Biology* 17 (2): 299–313.

Tufts University, Tufts Center for the Study of Drug Development. 2014. Cost to develop and win market approval for a new drug is $2.6 billion. *News*, November 18. http://csdd.tufts.edu/news/complete_story/pr_tufts_csdd_2014_cost_study.

Turner, L. and P. Knoepfler. 2016. Selling stem cells in the USA: Assessing the direct-to-consumer industry. *Cell Stem Cell* 19 (2): 1–4. www.cell.com/cell-stem-cell/flltex/S1934-5909(16)30157-6.

University of California, Hastings College of Law. California Ballot Measures. repository.uchastings.edu/ca_ballots/.

USA Today. 2004. Expand embryonic stem cell research. *USA Today Editorial*, June 14.

U.S. Centers for Disease Control and Prevention. 1992. The Fertility Clinic Success Rate and Certification Act. www.cdc.gov/art/nass/policy.html.

U.S. Congress. House of Representatives. Committee on Energy and Commerce. Stem Cell Research Enhancement Act 2005. 109th Cong, 1st sess. (H.R.810). https://www.govtrack.us/congress/bills/109/hr810.

U.S. Congress. House of Representatives. Committee on Energy and Commerce. Stem Cell Research Enhancement Act 2007. 110th Cong, 1st sess. (H.R. 3). https://www.govtrack.us/congress/bills/110/s5.

U.S Department of Commerce, Bureau of Economic Analysis. http://www.bea.gov/.

U.S. Department of Health and Human Services, Centers for Disease Control and Prevention, National Center for Health Statistics. 2015. Health United States. Table 19:107 and Table 14:93. www.cdc.gov/nchs/data/hus/hus15.pdf.

U.S. Department of Health and Human Services. Health Resources and Services Administration. 2014. U.S. Transplant Data by Center. http://bloodcell.transplant.hrsa.gov/research/transplant_data/us_tx_data/data_by_disease/national.aspx.

U.S. Department of Health and Human Services, National Institutes of Health. 2000. Notice; withdrawal of NIH Guidelines for Research Using Human Pluripotent Stem cells Derived from Human Embryos (published August 25, 2000, 65 FR 51976, corrected November 21, 2000, 65 FR69951). http://stemcells.nih.gov/staticresources/news/newsArchives/fr14no01-95.htm.

U.S. Department of Health and Human Services, National Institutes of Health. 2015. Estimates of Funding for Various Research, Condition, and Disease Categories (RCDC). https://report.nih.gov/categorical_spending.aspx.

U.S. Department of Health, Education and Welfare, Ethics Advisory Board. 1979.

U.S. Food and Drug Administration, Consumer Health Information. 2012. FDA Warns about Stem Cell Claims. www.fda.gov/downloads/forconsumers/consumerupdates/ucm286213.pdf.

U.S. Senate, Joint Economic Committee. 2000. The Benefits of Medical Research and the Role of the NIH. http://www.jec.senate.gov/public/index.cfm/republicans/2000/5/the-benefits-of-medical-research-and-the-role-of-the-nih.

Van Eperen, L. et al. 2010. Bridging the divide between science and Journalism. *Journal of Translational Medicine* 8: 25–27. https://translational-medicine.biomedcentral.com/articles/10.1186/1479-5876-8-25.

Vaughn, C. 2011. Stanford creates first PhD program in stem cell science. *Stanford Medicine News Center*, November 19. https://med.stanford.edu/news/all-news/2011/04/stanford-creates-first-phd-program-in-stem-cell-science.html.

Vaughn, C. 2012. Back in the Bay Area: Sabbatical draws stem cell expert Pedersen to Stanford. *Stanford Medicine News Center*, November 19.

Vedantam, S. 2004. Reagan's experience alters outlook for Alzheimer's patients. *Washington Post*, June 14. http://www.washingtonpost.com/wp-dyn/articles/A39072-2004Jun13.html.

ViaCyte, Inc. 2014. viacyte.com/clinical/clinical-trials/.

Victory, J. 2016. Journalists: 9 tips to combat stem cell hype in your news stories. *Health News Review*. http://www.healthnewsreview.org/2016/06/journalists-9-tips-to-combat-stem-cell-hype/.

Vierbuchen, T. et al. 2010. Direct conversion of fibroblasts to functional neurons by defined factors. *Nature* 463 (3): 1035–1041. http://www.ncbi.nlm.nih.gov/pmc/articles/PMC3552026/.

Vogel, G. 1999. Breakthrough of the year: Capturing the promise of youth. *Science Magazine* 286 (5448): 2238–2239. http://www.ncbi.nlm.nih.gov/pubmed/10636772.

Vogel, N. 2004. Investors pour millions into Prop. 71 race. *Los Angeles Times*, October 18. http://articles.latimes.com/2004/oct/18/local/me-prop71money18.

Voosen, P. 2011. Hiroshima and Nagasaki cast long shadow over radiation. *New York Times*, April 11. http://www.nytimes.com/gwire/2011/04/11/11greenwire-hiroshima-and-nagasaki-cast-long-shadows-over-99849.html?pagewanted=all.

Wade, N. 2001. Scientists divided on limit of federal stem cell money. *New York Times*, August 16.

Wade, N. 2005. Stem cell researchers feel the pull of the Golden State. *New York Times*, May 22. http://www.nytimes.com/2005/05/22/us/stem-cell-researchers-feel-the-pull-of-the-golden-state.html.

Wade N. and C. Sang-Hun. 2006. Human cloning was all faked, Koreans report. *New York Times*, January 10.

Wade, N. 2012. Stem cell work earns Nobel Prize. *New York Times*, October 8. http://www.nytimes.com/2012/10/09/health/research/cloning-and-stem-cell-discoveries-earn-nobel-prize-in-medicine.html.

Wade, N. 2013. The clone named Dolly. *New York Times*, October 14. http://www.nytimes.com/2013/10/14/booming/the-clone-named-dolly.html.

Wadman, M. 2010. U.S. stem-cell chaos felt abroad. *Nature* 467: 138–139. http://www.nature.com/news/2010/100907/full/467138a.html.

Waller-Wise, R. 2011. Umbilical cord blood information for childbirth educators. *The Journal of Perinatal Education* 20 (1): 54–60. http://www.ncbi.nlm.nih.gov/pmc/articles/PMC3209739/.

Warnock, M. A. 1985. *A Question of Life: The Warnock Report on Human Fertilisation and Embryology*. Oxford: Basil Blackwell.

Weiss, R. 1997. Scottish scientists clone adult sheep: Technique's use with humans is feared. *Washington Post*, February 24.

Weiss, R. 2001. Nobel Laureates Back Stem Cell Research, *Washington Post*, February 22. http://www.washingtonpost.com/wp-dyn/content/article/2005/08/02/AR2005080201092.html.

WETA. 2006. The Pros and Cons of IVF. The American Experience. https://www.pbs.org/wgbh/americanexperience/features/general-article/babies-pros-and-cons/.

WiCell Research Institute. 2016. University of Wisconsin, Madison. http://www.wicell.org/home.

Wilke, J. and L. Saad. 2012. Older Americans' moral attitudes changing. *Gallup Polls*. www.gallup.com/poll/162881/older-american-moral-attitudes-changing.aspx.

Williams, R. 2008. Sir John Gurdon: Godfather of cloning. *Journal of Cell Biology* 181 (2): 178–179.

Wolfe, A. 2016. Susan L. Solomon's stem-cell research quest. *Wall Street Journal*. http://www.wsj.com/articles/susan-l-solomons-stem-cell-research-quest-1454699397.

Yamanaka, S. 2012. Induced pluripotent stem cells: Past, present, and future. *Cell Stem Cell* 10 (6): 678–684. http://www.ncbi.nlm.nih.gov/pubmed/22704507.

Yamanaka, S. 2013. Stem cells in translation. *Cell* 153 (6): 1177–1179. http://www.cell.com/cell/issue?pii=S0092-8674(13)X0012-1.

Yong, E. 2016. Testing drugs on mini-yous, grown in a dish. *The Atlantic*, June 22. http://www.theatlantic.com/science/archive/2016/06/testing-drugs-on-mini-yous-grown-in-a-dish/488039/.

Yu, J. et al. 2007. Induced pluripotent stem cell lines derived from human somatic cells. *Science* 318 (5858): 1917–1920. http://www.ncbi.nlm.nih.gov/pubmed/18029452.

Zitner, A. 2001. Uncertainty is thwarting stem cell researchers. *Los Angeles Times*, July 16. http://articles.latimes.com/2001/jul/16/news/mn-22904.

Index

Note: Page numbers followed by "*n*" indicate notes.

E

F

G

H

I

J

K

L

M

N

O

P

Y

Z